PREVAIL

Reclaiming the Divinity of our Humanity

By Dr. Noah Manyika

A project of

.the
issachar
coll3ctive
ENVISION · BUILD · TRANSFORM

Prevail

Reclaiming the Divinity of our Humanity

Published by The Issachar Coll3ctive

ISBN: 978-1-965649-14-5

Copyright © 2025 by Dr. Noah Manyika

Cover design by The Issachar Coll3ctive

Interior design by atritex.com

Available in print from your local bookstore, online, or from the publisher.

For more information on this book and the author visit: noahmanyika.com

Library of Congress Cataloging-in-Publication Data

Manyika, Noah

Prevail: Reclaiming the Divinity of our Humanity / Dr. Noah Manyika, 1st ed.

Printed in the United States of America

Contents

Author Testimonials

"When Noah Manyika speaks, it's hard not to be transfixed."

Mary Wong - *President*
Office Depot Foundation

"One of the most gifted communicators I know... rare deep intellect and ability to connect with audiences of all types."

Jay Hein – *CEO*
Sagamore Institute

"A leader of leaders in a class of his own...inspiring master coach with unique abilities to educate and dissect micro and macro issues on a global scale."

Manny Ohonme – *Founder/CEO*
Samaritan's Feet & The World Shoe

"An outstanding orator and brilliant thought leader on global issues and trends. I have heard him hold many audiences in Washington, D.C. spellbound."

Dr. Allan Goodman - *President*
International Institute of Education (IIE)

Reviews of Previous Published Work

Reviews of "The Challenge of Leadership: Is There Not a Cause?"

"This book has the inspirational edge of a Dale Carnegie piece and the moral and spiritual backbone of the book of James. It is a sermon on leadership, not a theoretical, ivory tower polemic. It is a must read for anyone attempting to be a leader and for those who sense the need to lead but lack the courage to do so!"
Julian Champion – *Founder*
West Point School of Music, Chicago Il.

"An excellent work…one of the two best I have ever read on Christian leadership. Noah thank you for enduring the process and writing of it."
Steve Martin - *Director*
Derek Prince Ministries.

"Powerful work…Noah has offered superb clarity and discernment on the way God regards the leadership issue, both in Christian and secular circles."

Dennis Matangira – *Senior Managing Partner*
Zebu Investment Partners.

"Powerful reading. At a time when the church desperately needs leaders at all levels, Noah Manyika, with much insight and wisdom, has written a critically important book. It is extremely biblical. It is practical. It is real. It will change the lives of those who have a calling to spiritual leadership."

Byron Wicker – *Senior Pastor*
RiverLife Fellowship, NC.

"With the applied logic, craftsmanship and inspired passion reminiscent of the classical artisan and follower of Christ who crafted the New Testament argument that Christ was the promised Messiah, Manyika challenges us to make the essential connection between unconditional submission and mission, and to rise to the virtue of availability."

Dr. Jameson Kurasha – *Professor*
Religion, Classics and Philosophy

Reviews of "Redeeming Sundar: Faith and Innovation in the Age of AI."

"With the digital world advancing at such a rapid pace, *Redeeming Sundar* is a masterful and powerful call to action, inspiring believers to lead with wisdom, creativity, and unwavering commitment to Kingdom values."

DeAngelo Burse - *Professor,*
TEDx and Keynote Speaker & Author

"The majority of the church is living in denial, settling into a worrisome irrelevance that separates faith from science. It is time for the church to heed Jesus' brutal critique that the children of this world are wiser in their generation than the sons of light. If we are to engage the world, we need language for believability - and this is what this powerful book generously provides."

Dr. Oladotun Reju - *Professor*
Transformational Leadership.
Bakke Graduate University, Dallas, TX.

"With profound insight and clarity, Dr. Manyika masterfully bridges the worlds of faith and innovation, challenging readers to see technology not as a threat, but as a powerful tool for advancing the Kingdom of God. *Redeeming Sundar* is an essential guide and must read for anyone with a passion to see the flourishing of God's Kingdom in an ever-changing world."

Rev. Tapiwa Nduna – *Founder*
Bread of Life International Ministries, UK.

"As a Christian leader in international development, I am deeply challenged by Noah's call to integrate faith and innovation. This book offers Biblically rooted actionable insights for effecting socioeconomically transformative faith-driven innovation. We can no longer separate technological progress from our spiritual mandate to transform lives."

Dr. Komborero Choga – *Senior Director*
Compassion International.

"The men of Issachar were known for their wisdom and understanding of the times. Noah Manyika is one of them. Disruptive technology has been a constant challenge to navigate for faith leaders throughout history. *Redeeming Sundar* will profoundly inspire you to embrace emerging technologies while remaining grounded in the fundamental Great Commission mandate."

Rev. Michelle Hoverson – *Founder*
Lake Norman CDC, NC, USA.

"Dr. Manyika boldly and compellingly advocates for integrating faith and innovation from a biblically grounded perspective. His unique insights draw from a remarkable multidisciplinary journey - including life behind the Iron Curtain in Romania, studies at Georgetown under leaders like Madeleine Albright, advanced degrees from Bakke and Vision universities, and diverse experiences as pastor, missionary, entrepreneur, CEO, and 2018 Zimbabwe Presidential Candidate. This book is essential reading for anyone committed to bridging faith and technology for Kingdom advancement and human flourishing."

Dr. Gwen Dewey – *ex President*
Bakke Graduate University, Dallas, TX

"This work is more than a polemic on technology and its usage. It is a call to arms for the church to rise up a little higher and seize the opportunities before us to use all means possible to preach the gospel of the Kingdom of God to the four corners of the globe. I highly recommend the book you find in your hands and the overall ministry of Dr. Noah Manyika, a visionary and apostle for our time and season."

Dr. Stan DeKoven – *President*
Vision International University, USA.

"*Redeeming Sundar* is an outstanding and exceptionally well-written book. It provides innovative, practical insights and effective strategies for overcoming internal and external obstacles in navigating a world where rapidly accelerating technology reshapes our lives in ways that once seemed purely futuristic."

Byron Wicker – *Senior Pastor*
RiverLife Fellowship, Mooresville, NC.

"I've known Noah for more than twenty years. Several things remain consistent with him. He lives and breathes leadership. He tries to understand the geopolitics surrounding us today. He is amazingly innovative. And he wants to find out new ways to reach and serve people for the Lord Jesus who guides his life. With *Redeeming Sundar* he proves once again that his work is always worth reading. You'll certainly know more and be better because of it."

David Chadwick - *Senior Pastor*
Moments of Hope Church, Charlotte, NC.

PREVAIL

Reclaiming the Divinity of our Humanity

Acknowledgements

More than four decades ago, in the wood-paneled stillness of his office in Dedinje, Belgrade in the former Yugoslavia, my father - an ambassador navigating the crosscurrents of Cold War diplomacy - turned to me with a thoughtful look and asked for my opinion.

I was studying in Bucharest at the Ştefan Gheorghiu Academy, a school devoted to the ideological formation of future Romanian and "Third World" leaders. Eager to share what I had learned, I launched into a confident explanation of dialectical materialism and scientific socialism - drawing from the lectures of my professors and the debates I'd had with fellow students from Syria, Cyprus, Tunisia, Palestine, and Namibia.

My father listened attentively, head slightly tilted, absorbing every word. When I had finished, he paused, then asked with a quiet clarity that gently dismantled all my rhetoric:

"What does all that really mean?"

It was not a dismissal - it was a gift. A quiet, piercing invitation to search for clarity, to distinguish the essence from the echo, the truth from its trappings. That question has stayed with me - not only as a challenge to ideological complexity, but as a spiritual compass, reminding me that language should never outpace meaning, and theology should never be divorced from life.

My father looked at things through the clear eye of faith. In addition to Yugoslavia, he was also accredited to Enver Hoxha's Albania, one of the harshest, most repressive Marxist-Leninist regimes in the world - an iron-fisted dictatorship masquerading as a workers' paradise. There, peasants, workers, and ordinary citizens lived as far from a utopia as the characters in George Orwell's 1984, trapped in a grim and all-consuming nightmare.

None of these places resembled the promised "dictatorship of the proletariat" so triumphantly heralded by Marxist-Leninist propaganda, where the needs of the worker supposedly reigned supreme. Instead, they were brutal autocracies, ruled by greedy, rapacious and insecure elites who preyed upon the very people they claimed to liberate - citizens weighed down by the crushing burden of ideology, reduced to chanting hollow praise songs to their rulers in desperate exchange for the rations they needed to survive the nightmare they were told was a revolution.

In that quiet conversation, my father had also caught me in an act of hypocrisy. He knew from my own accounts how dire the situation was in Ceaușescu's Romania, where I was studying: the food shortages; the babies dying in hospital incubators because of power cuts; the six-month bans on private driving due to chronic fuel shortages. And yet, even amid these daily hardships, staged rallies echoed with chants like,

"Ceaușescu și poporul! Ceaușescu-România, mândria și onoarea noastră!" ("Ceausescu - Romania, our pride and esteem!")

The newspapers faithfully carried the illusion. In *Scînteia* – the Romanian Communist Party's official mouthpiece - reports of these "extraordinary" rallies were printed verbatim, with the editors meticulously recording the nature of the crowd's applause at every stage of the speech: "moderate," "powerful," "mild," as if revolution itself could be measured by the intensity of orchestrated ovations.

The theories taught at places like the Ștefan Gheorghiu Academy in Bucharest - a training ground for Romanian Communist Party cadres and a site for indoctrinating foreign students - or at institutions like Patrice Lumumba University in Moscow or the Karl Marx Higher School of the Socialist Unity Party in East Berlin, were a far cry from the cruel realities on the ground.

I thank my father, Kennedy Grant Dick Manyika, not only for that question, but for a life marked by grounded faith and uncommon wisdom. And I honor my mother, Rahab Manyika, who walked that faith journey beside him - believing with him for the miracle of healing and deliverance from infirmity and bondage, and standing with him as he rose to the heights of diplomatic service and church leadership in the land of their birth.

Together, they modeled a faith powerful in its simplicity - not loud, but weighty. Their lives embodied what this book seeks to illuminate: that the divinity within us - our essence as image-bearers of God - is not theological theory, but daily truth waiting to be reclaimed.

I dedicate this book to their memory.

And to all who seek to make sense of life beyond dogma - to those disillusioned by institutional religion and propaganda and empty self-praise yet hungry for presence, power, and purpose - this is an offering. Not from one who has apprehended all things, but from one who, like Paul, presses on:

"Brethren, I do not count myself to have apprehended; but one thing I do... I press toward the goal for the prize of the upward call of God in Christ Jesus." (Philippians 3:13-14)

Faith does call us to seek deep things - but not in pride. As David wrote:

"Lord, my heart is not haughty... Neither do I concern myself with great matters, nor with things too profound for me... Like a weaned child is my soul within me." (Psalm 131)

There are deep truths. But they begin in simplicity:
- In believing what God has spoken - even when others demand more proof.
- In living what we have received - even when we do not understand everything.

- In choosing the unadulterated word of God over high-sounding philosophies - even those preached from pulpits - that obscure what He has made plain.

For in God's original word, we find not confusion, but clarity; not burden, but blessing; not bondage, but the birthright of dominion and purpose. The truth, simply stated and never needing to be complicated, is this: God fulfilled His creative intent:

> "Then God said, 'Let Us make man in Our image, according to Our likeness; let them have dominion over the fish of the sea, over the birds of the air, and over the cattle, over all the earth and over every creeping thing that creeps on the earth.' So God created man in His own image; in the image of God He created him; male and female He created them. Then God blessed them, and God said to them, 'Be fruitful and multiply; fill the earth and subdue it; have dominion over the fish of the sea, over the birds of the air, and over every living thing that moves on the earth.'" (Genesis 1:26–28)

We are God's finished workmanship from Genesis - *fearfully and wonderfully made*, as the psalmist sings (Psalm 139:14) - and *blessed by God* to fulfill our mandate.

The Fall was not the end of the story. The finished work of the Cross stands as the bridge back to our original design. Our responsibility is to live out what is finished.

Our charge is to prevail.

I am grateful to Dr. Bill Payne, Director of Dissertations and Final Projects at Bakke Graduate University, who supervised the research that formed the foundation of my Doctor of Transformational Leadership degree and from which the *Prevail* model emerged. I also extend my thanks to Dr. Gwendolyn Dewey and Dr. Martine Audéoud, who served on the dissertation committee and the final review team. Special thanks to Eddie Jones (writerscoach.us), whose steady guidance helped bring this manuscript - and that of my previous book, *Redeeming Sundar: Faith and Innovation in the Age of AI* - to completion with care and clarity.

And to my wife, Phillis Manyika (née Mugabe) - thank you for your grounding wisdom and sharp insights. Your partnership is a quiet strength woven through every page of this journey.

May this book point others to the essence of their calling, the weight of their presence, and the divinity of their humanity. May it call forth not just understanding - but awakening.

Dr. Noah Manyika
Charlotte, NC, USA
April 2025

There are no ordinary people.
You have never talked to a
mere mortal. Nations, cultures, arts,
civilization - these are
mortal, and their life is to ours as
the life of a gnat. But it is
immortals whom we joke with, work
with, marry, snub, and
exploit - immortal horrors or
everlasting splendors."
C.S. Lewis, The Weight of Glory

PROLOGUE

"Deep calleth unto deep"
Psalm 42:7

What if the deepest truth about you isn't where you were born, to whom you were born, your race, your gender, or your struggles, but an eternal essence crafted before time, in the very image of God?

What if - as a parent, pastor, educator, global Christian development leader, community leader, theologian, or coach - awakening the divinity within the humanity of those you are raising, serving, and leading becomes the defining purpose of your life, service, and vocation? "Deep calleth unto deep," the Psalmist sings (Psalm 42:7). This book seeks to help us embrace the higher calling of living, parenting and leading from our depth.

It starts with a long-overdue theological correction, returning us to the critical place of beginnings where our journey is guided once again by God's blueprint. The wisdom and benefit of "returning" are articulated profoundly by the preacher in Ecclesiastes 9:11:

> "I returned and saw under the sun that –
> The race *is* not to the swift,
> Nor the battle to the strong,
> Nor bread to the wise,
> Nor riches to men of understanding,
> Nor favor to men of skill;
> But time and chance happen to them all."

Seeing, or understanding here, is associated with returning - going back from the conclusion that life favors some and not others; that God condemned us to permanent suffering because we were born in places defined by misery instead of human flourishing; that life is permanently fixed. This view misinterprets Paul's teaching in 1 Timothy 6:6 ("Now godliness with contentment is great gain"), incorrectly implying we must accept servitude as our predetermined fate. Paul indeed warns about being driven by a "desire to be rich" and "the love of money," which ensnares many into harmful lusts (vv. 9-10).

Strong words indeed. Jesus does remind us in John 10:35 that "The scripture cannot be broken," that God's truths only exist in tension because we tend to recruit one truth to support a personal choice that seems to be contradicted by another. God's blueprint- clearly creating man with the capacity and permission to have dominion, be fruitful, multiply, and master life (Genesis 1) - is hardly hostile to Paul's warnings. The desire to master life for redemptive purposes - feeding our families, ensuring quality education, providing medical care, leaving an inheritance (Proverbs 13:22), and supporting righteous leadership (Proverbs 29:2) - is not evidence of greed or the love of money.

Accepting the truth in Ecclesiastes 9:11 - "But time and chance happen to them all" - means embracing humanity's capacity and permission to prevail, as exemplified by David prevailing against Goliath; by the once-mocked nerds and autistic kids now leading the technology world and without whose innovation the "normal" people and "gifted" jocks who once tormented them would be jobless; by diminutive NBA players Muggsy Bogues and Spud Webb, who outplayed and outlasted many "favored" with the gift of height in the most competitive basketball league in the world; and by Stevie Wonder's lasting musical genius despite blindness.

Recognizing the depth God gave humanity at creation - a foundation of dignity and vitality - cannot ignore the reality of human suffering. Edwin Markham's poem, "The Man with the Hoe" (1898), vividly depicts how exploitative social systems strip humanity of dignity, rendering man unrecognizable from the beings God created in His image and likeness:

> *Is this the Thing the Lord God **made** and **gave***
> *To have dominion over sea and land;*
> *To trace the stars and search the heavens for power;*
> *To feel the passion of Eternity?*
> *Is this the Dream He dreamed who shaped the suns*
> *And marked their ways upon the ancient deep?*

Markham emphasizes that human reduction through suffering prevents man from fulfilling his divine purpose. He can't have dominion over the sea and land. He *can't trace the stars and search the heavens for power.* He can't *feel the passion of Eternity.* His reduced state mocks the truth of God's word that man is indeed "the Dream He (God) dreamed," the special one of all that He created so worthy of His love He gave "His only begotten

Son, that whoever believes in Him should not per-
ish but have everlasting life." (John 3:16).

His poem is a powerful critique of oppression
and a profound call to leadership:

O masters, lords and rulers in all lands,
Is this the handiwork you give to God,
This monstrous thing distorted and soul-quenched?
How will you ever straighten up this shape;
Touch it again with immortality;
Give back the upward looking and the light;
Rebuild in it the music and the dream,
Make right the immemorial infamies,
Perfidious wrongs, immedicable woes?

The implication is clear: the people we raise,
lead, coach, and teach represent the handiwork
we offer to God. We can model the redemptive
impact we are called to have by following Jesus'
example articulated in John 17, where He com-
mends the disciples to God in prayer. It begins
with recognizing that those we have responsibility
for - whether children, church members, or those
served through community or global development
programs - belong to God. Our responsibility is to
shape them according to God's desire:

- "I have manifested Your name to whom You have given Me out of the world. They were Yours, You gave them to Me, and they have kept Your word." (v.6)
- "Now they have known that all things which You have given Me are from You. (v.7)
- "For I have given to them the words which You have given Me; and they have received them, and have known surely that I came forth from You; and they have believed that You sent Me."(v.8)
- I do not pray that You should take them out of the world, but that You should keep them from the evil one." (v.15).
- Sanctify them by Your truth. Your word is truth. (v.17)
- As You sent Me into the world, I also have sent them into the world." (v.18)

Markham challenges us to reflect on the handiwork our parenting, Christian leadership, training, and coaching present to God. He is not merely concerned with appearance or what the handiwork will look like. His writing probes a deeper truth: What will the people who pass through our hands truly become? Will they embody in their essence "the Thing the Lord God made," or remain the diminished "stolid and stunned" creatures, reduced, "profaned, and disinherited" by life? Markham's questions to the masters, lords, and rulers - "How will you ever straighten up this shape" or "touch it again with immortality" - highlight a deeper responsibility: not merely to mold others but to awaken the divinity within their humanity.

Humanity's crisis begins when, for whatever reason, we lose sight of our true identity - our essence. While it is tempting to blame exclusively external factors and oppressive systems that reduce individuals to mere "slaves of the wheel of labor," consigning them to perpetual servitude benefiting a privileged few, our condition also arises from individual and collective cultural or religious mindsets. These mindsets mistakenly equate our essence with our fleshly identity and social misfortunes, causing us to accept reduction as inevitable. We remain trapped in a swamp from which the evil one, who seeks to prevent our recognition of God's divine design, recruits us to wage war against ourselves.

Markham confidently calls his warning to leaders a prophecy, declaring that judgment, manifesting as "whirlwinds of rebellion," is inevitable because humans will ultimately rebel against their imposed diminishment:

How will it be with kingdoms and with kings -
With those who shaped him to the thing he is -
When this dumb Terror shall rise to judge the world.
After the silence of the centuries?

Being diminished does not mean people will not resist oppression. The critical questions are how they fight, who they become in the struggle, their motivations, and what they ultimately seek to restore. History has repeatedly shown that rebellions waged by diminished people often fail to produce lasting beneficial results. As Hosea 8:7 warns, "they sow the wind, and they shall reap the whirlwind." When the "dumb Terror" rises to judge the world without first reclaiming humanity's divinity - which includes not only the capacity for dominion but also the accompanying redemptive responsibility - it unleashes judgment not only on former oppressors but also upon itself. The suffering continues, only under a new set of oppressors.

The premise of this book is that genuine transformation - of homes, communities, and countries

- begins with restoring who we truly are as God's handiwork. I use the present tense "are" intentionally because the divinity is very much within us, though often entombed in our flesh rather than embodied. Like Lazarus, it is ready to respond to the call to come forth from within those who are "His workmanship, created in Christ Jesus for good works, which God prepared beforehand that we should walk in them." (Ephesians 2:10)

It is always tempting to rush toward DOING without giving sufficient attention to BEING. Before defining the "good works" we were created for in terms of ministry programs, we must recognize that Satan's primary battle is to ensure we do not understand our true identity - the essence of who God made us to be. The Jews who wanted to stone Jesus in John 10 clearly articulated their motive: not because of His good works, but because He declared who He was. However, their challenge was not merely His divine claim but also His reminder of their own divine identity, as seen in their reaction when He quoted scripture:

"Is it not written in your law, 'I said, "You are gods"'? If He called them gods, to whom the word of God came (and the Scripture cannot be broken), do you say of Him whom the Father sanctified and sent into the world, 'You are blaspheming,' because I said, 'I am the Son of God'?" (vv.34-36)

Who He *was*, represented a greater threat than His actions. Similarly, our BEING - created in God's image and likeness (Genesis 1) - is a more significant threat to the kingdom of darkness than our actions alone. This does not imply our actions are irrelevant; however, if Satan can diminish our sense of identity (our BEING), he can limit the impact of our actions, since the true impact of kingdom work flows from our divine essence.

The Samaritan woman's confusion in John 4, which we will explore in greater detail in Chapter 4, illustrates the necessity of reclaiming true identity before meaningful action can occur. Initially, she speaks proudly of her Samaritan heritage, unaware that this identity traps her in misunderstanding. Jesus refuses to interact with her based on the Jewish identity she expects, because that identity had hindered the Jews themselves. Awaiting the Messiah, they ultimately rejected and crucified Him, unable to escape the limitations the self-imposed identity that fought against their best interests. Jesus' insistence on a deeper conversation brought to life the Samaritan woman's true essence, previously buried under her cultural identity. Only after reclaiming her true BEING could her impactful DOING begin.

However, it is critical not to imply that the only impact of reclaiming the divinity of our humanity should be to turn people into powerful "street preachers" like the Samaritan woman whose testimony brought many to Christ. The good works that flow from our true identity must indeed point the way to Christ, but not just in a limited religious way. It must validate the whole truth that He represents and cultivate the fullness of life that He, according to John 10:10 came to cultivate.

All truth includes, indeed begins with, the understanding that we were created to master life. Voltaire's assertion, "God gave us the gift of life; it is up to us to give ourselves the gift of living well," may challenge those who feel trapped by circumstances beyond their control. Yet, it encapsulates a vital truth that even the oppressed urgently need: they possess inherent dignity, capacity, and divine permission to shape their lives. True liberation begins when individuals recognize their intrinsic worth and actively engage in their journey toward freedom, often aided by the "helpers of the war" (1 Chronicles 12:1) God inevitably raises.

The goal of this work is to equip believers to participate in and lead God's redemptive work on earth with clarity and confidence, anchored in the indisputable truth that identity and dominion - according to God's declaration of creative intent in Genesis 1 - form an indivisible unity. Although the concept of human divinity might trigger reflexive resistance from self-appointed guardians of God's divinity, the following points are indisputable:

1. God's explicit intention to share His image and likeness (Genesis 1:26).
2. God's active fulfillment of this intention in creating humanity (Genesis 1:27–28).

God did not share His divine position or the fullness of His nature. Nevertheless, the measure of Himself He imparted, coupled with permission to exercise dominion, is sufficient to define our essence and fulfill our redemptive purpose - here on earth, not in heaven. As Jesus affirmed in His prayer in John 17:15, our assignment is earthly; He did not ask the Father to remove the disciples from the world.

This book is founded on the belief that no one is created helpless or hopeless; all are created and empowered to prevail here. Humanity's divine identity is foundational, providing inherent dignity, capacity, and divine permission to master life. The greatest threat to the kingdom of darkness is the one who awakens the divinity of humanity - both their own and that of those they lead -draining the swamp from which the enemy recruits us to wage war against ourselves.

We see this truth mirrored in history. Holocaust survivors returning to Israel chose resilience over resignation despite horrific experiences. Shimon Peres, in the foreword to *Start-up Nation: The Story of Israel's Economic Miracle*, expressed this powerfully:

> "The seeds of a new Israel grew from the imagination of an exiled people... The exile was extremely long, some two thousand years. The exiles left the Jews with a prayer and without a country... With the establishment of the State of Israel, this great prayer was planted in a small land. The soil was obstinate, and the environment was hostile... We had to create ourselves anew. As a poor people coming home to

a poor land, we had to discover the riches of scarcity. The only capital at our disposal was human capital. The kibbutz became an incubator, and the farmer a scientist... Israel bred creativity proportionate not to the size of our country, but to the dangers we faced."

Without reclaiming the divinity of their humanity and the capacity and permission to prevail it implied, it would have been impossible to turn deserts into flourishing gardens and kibbutzim into the amazing incubators of innovation that have made Israel the ultimate "Start-Up Nation," representing, according to Senor and Singer, "the greatest concentration of innovation and entrepreneurship in the world today."

Ultimately, this work introduces *PREVAIL* - an acronym for **P**resence, **R**elevance, **E**xploring, **V**alidating, **A**lleviating, **I**nspiring, and **L**everaging. These seven principles form a structured framework for impactful living and the fulfillment of humanity's redemptive purpose. By reclaiming the divinity of our humanity, we equip ourselves and others to transcend suffering, embrace our true identity, and pursue transformation that lasts.

At the end of each chapter, you'll find a *Leader and Kingdom Citizen Blueprint* - a practical guide designed to help you embody each principle in your roles as a leader, disciple, parent, colleague, or citizen. These blueprints:

- Offer actionable leadership and kingdom citizenship guidance,
- Translate theology into everyday decision-making,
- Empower you to lead from identity, not just position.

To further support your journey, you'll also find a collection of *Prevail Scorecards* at the end of the book (Part Two). These are not exams, but invitations - spaces for reflection, alignment, and growth. Each scorecard corresponds to a chapter and is designed to:

- Help you assess your current alignment with the principle,
- Prompt honest self-examination and insight,
- Encourage intentional steps toward becoming who you were created to be.

We begin now where all true leadership and redemptive kingdom citizenship begins - not with doing, but with being.

Gravitas

We were not formed in haste or whim
But with design, both deep and vast:
Gravitas – the weight of being,
The gift, the meaning, the power to live.
In every soul, the echo rings,
Of purpose bound to The Holy One.

Noah Manyika - 2025

CHAPTER 1
Presence: Making the Weight

Dominion belongs to those
who make the weight.

Weighing In

In the sport of boxing, no one enters the ring without first stepping on a scale. Fighters must **make the weight** - meeting the exact requirements of their designated weight class - to be eligible to compete. You cannot fight out of your class. You cannot bypass the scale. You cannot enter the arena unless you've proven that you belong in it.

In the Kingdom, the principle is no different. To show up with authority, you must first carry the weight of identity. Being created in the image and likeness of God is the weight class to which every human being belongs. It is not a title earned by spiritual performance or religious affiliation, but a divine reality that precedes both. **Imago Dei** is not lightweight.

Presence, in the biblical sense, is not passive. It is weighty. And every assignment in life - whether in the marketplace, the home, or the public square - demands that we **make the weight** before we try to throw a punch.

Being

Before there was doing, there was Being. In the beginning, God WAS - not a mere ambiance, not an ethereal abstraction, but an eternal, weighty Presence that precedes and transcends all action.

Before the heavens stretched wide or the earth took shape, there was I Am -the God who simply is. And when Moses stood before the burning bush, barefoot and bewildered, hesitant and unsure, asking for the name of the One who was sending him into Pharaoh's court, God replied:

"I Am who I Am."

Not *I do what I do.*
Not *I will accomplish what I will accomplish.*
Simply I Am.

This was no philosophical riddle – it was a statement of fact. God does not root authority in activity, but in identity. Identity brings weight to being. We were created in the image and likeness of God. That is the weight of our being – the substance, the gravitas of our existence.

In *Prevail*, Presence is defined as the weight of being. In Scripture, the word Presence is often used as a stand-in for God's personal name or identity, functioning as a proper or substantive noun. In other words, God is Presence. We were created in the image and likeness of Presence, so whatever Presence implies cannot be separated from our being, our identity.

The Weight of Presence

The word **"Glory"** in Scripture is translated from the Hebrew word **kabowd**, derived from the root דָּבַכ (**kaved**), meaning "to be heavy" or "to be honored." According to Strong's Lexicon, it describes the weightiness or significance of something or someone, often in terms of reputation, wealth, or splendor. In the context of God, **kabowd** refers specifically to His majestic presence and the manifestation of His divine attributes.

God's glory is inseparable from His being. He is not God without the weightiness of His presence. Likewise, we cannot truly bear His image without carrying the weight inherent in the divinity of our humanity. It's important to regularly step onto God's spiritual scale - to evaluate whether we are bringing the full weight of our being into every situation we encounter. Most of our encounters will fall into the heavyweight class. We were created in the image and likeness of God - because no one who can't make the weight belongs in the ring.

Among the seven dimensions of impact in the *Prevail* model (Presence, Relevance, Exploring, Validating, Alleviating, Inspiring, and Leveraging) Presence is foundational. Genuine presence - reflecting the essence of our being - establishes our relevance. It fuels our courage to explore. It provides confidence to validate truths we hold. It enables us to see others as God sees them, prompting us to alleviate pain and suffering in the most God-honoring ways. Presence empowers us not merely to inspire - to genuinely breathe life into others - but also to leverage whatever God places in our hands to advance His Kingdom.

Therefore, you'll see the remaining principles in *Prevail* articulated through Presence: Relevant Presence, Exploring Presence, Validating Presence, and so forth. Presence is powerful, even without action. The Psalmist, celebrating the delightful weightiness of God's presence, sings:

"In Your presence is fullness of joy; at Your right hand are pleasures forevermore." (Psalm 16:11)

Moses understood this profound power of presence. As he led Israel from the chains of Egypt into the uncertain terrain of promise, he emphatically declared: "If Your Presence does not go with us, do not carry us up from here." (Exodus 33:15) For the children of Israel, Presence was not simply helpful - it was essential. More than strategy or resources, it was Presence that they most needed on their journey toward the Promised Land.

The Power of Presence

In Psalm 97:5, the Psalmist captures a profound truth: "The mountains melt like wax at the presence of the Lord." This is no mere poetic flourish. It's a vivid depiction of presence as a **force** - not passive or ornamental, but active and atmosphere-altering.

Consider the immense energy unleashed by radiation during a nuclear meltdown. It's unseen yet potent, capable of influencing environments and altering lives hundreds, even thousands, of miles away. It doesn't announce itself. It doesn't require movement. And yet, its presence is undeniable. It permeates everything in its path, silently transforming the very fabric of what surrounds it.

Now picture something even quieter - **a pound of enriched uranium sitting unnoticed on a sidewalk.** It doesn't explode. It doesn't shout. But it still has the power to alter DNA, compromise infrastructure, and endanger lives. It doesn't have to do anything. It just has **to be.**

So it is with those who carry the weight of divine assignment. Like enriched uranium, a person sent by God doesn't need a platform to make an impact. Their mere presence leaves a mark - on hearts, institutions, atmospheres, and history itself.

Radiation is unseen but profoundly present. It doesn't ask permission to affect its environment. Its very nature changes things. And that is precisely the nature of radiant presence: the kind of presence that stems from divine calling and identity.

Radioactivity becomes a powerful image for image-bearing presence: invisible, weighty, and impossible to ignore. Presence like this doesn't speak first - it registers. It doesn't persuade - it radiates. When God sends someone into a space filled with disgrace or danger, their authority isn't in their volume - it's in their essence.

But unlike harmful radiation, the presence of God is infinitely benevolent. It doesn't destroy - it restores. While nuclear radiation corrupts what it touches, **God's presence heals, renews, and transforms.** Just as radioactive particles penetrate walls and skin, God's presence effortlessly reaches into the most fortified and hidden places of our hearts, reshaping realities and awakening what was dead.

Psalm 68:8 and Psalm 114:7 describe the earth itself trembling and shaking before the Lord - another powerful metaphor of how tangible and impactful His presence truly is. The presence of God radiates goodness, dissolving barriers, melting obstacles, and reshaping landscapes of despair into territories of hope and purpose.

Indeed, the Psalmist affirms in Psalm 73:28, "The nearness of God is my good." Unlike the dangerous proximity to nuclear radiation, closeness to God's presence is life-giving. The nearer we come to Him, the more profoundly we experience healing, empowerment, and wholeness. His presence doesn't weaken or destroy - it strengthens and restores. It ignites within us an extraordinary power, unlocking our capacity to prevail, to rise, and to fulfill the redemptive potential deeply embedded within us.

This is the radiant power of presence: transformative, restorative, and infinitely good. It is the essence of the divine encounter, turning mere existence into vibrant living, equipping us to prevail in every sphere of our lives.

A Radiant Gift from God

We come into the world from God's radiant presence. This is a profound truth because of where we were created before we were formed in our mother's wombs. We come from God (Psalm 127:3) as radiant gifts to the world. We don't become radiant by calling. We were created radiant. Part of the reason for problems the faith community is experiencing is because we have it in reverse. People don't become better kingdom citizens because they have become leaders. They become better leaders because they are better kingdom citizens, trained up at home, not by seminaries.

At the heart of how we re-enter the spaces from which the faith community has withdrawn - those places that shape how society functions - is the recovery of the radiance we lost as children of God, not even as leaders. Resistance to our late-stage attempts to re-enter the civic square, where the rules of engagement have already been established by others, is predictably fierce, and our efforts often inflame the toxicity of our times. What we need most is the recovery of Presence and a renewed understanding of our identity - an understanding we ironically lost through religion itself.

Before we can innovate, influence culture, or redeem our marketplace assignments, we must return to the **ground of being.** That is first and foremost a theological step. We must recover a settled knowing - an identity-rooted confidence that precedes performance. It is a knowing that makes us unshakable, not because we are perfect, but because we are anchored.

The life of a Polish woman who lived more than a century ago offers a powerful metaphor for the **radiancy of being** that is critical to fulfilling our redemptive purpose. The compelling lesson from the life of the girl born to Bronisława and Władysław Skłodowski in Warszawa, Poland, lies not only in her immense contributions to the science of radioactivity, but in the evident settledness of her awareness that she was more than merely a girl - she was a **radiant gift from God.** Her name was Maria Salomea Skłodowska, known to history as Marie Curie.

Born into a family whose fortunes were stripped away for resisting the oppression of the Russian Empire, Maria came of age in a world that constantly told her she couldn't - especially because she was a girl. Denied access to university due to her gender, she enrolled in the *Uniwersytet Latający* - the "Flying University," an underground network of Polish dissidents committed to providing education, including to women, outside the reach of state ideology. She thrived there, and went on to become not only the first woman to win a Nobel Prize, but the first person to win it in two distinct scientific disciplines.

The world remembers what she did - but what mattered more was who she was. No one can consistently achieve what they have not first become. It was Marie Curie who coined the term *radioactivity.* Together with her husband, Pierre, she discovered radioactive elements such as Polonium and Radium.

Marie Curie was not a religious woman; historians note that her choice to wear a dark blue outfit instead of a traditional bridal gown at her wedding was a deliberate protest against church-sanctioned sexism. Religion, it seems, failed to recognize her as the radiant gift she was.

Though she may never be canonized by religious institutions, Marie Curie grasped a truth many religious people of her time did not: she was more than *just* a woman. She boldly embodied her full humanity, defying gender, religious, political, and societal limitations simply by being authentically herself. That authentic presence preceded every world-changing achievement - from naming radioactivity and discovering new elements, to revolutionizing medicine with imaging techniques and radiotherapy. Ultimately, her courageous embodiment of her essence paved the way for breakthroughs that continue to save lives and inspire generations.

And by the way - Marie's quest was never to become a leader. She simply wanted to be Marie. Just as when David arrived at the Valley of Elah and challenged the cowardice of Saul's army, he wasn't seeking to be recruited into military service. He was simply being David.

The Need for a Settled Knowing

Believers are not in short supply. You'll find us in every sphere of society, including the marketplace. But this book is not a call for more visible religiosity. There can be no greater witness to the Way (Jesus), no more compelling validation of the truth of the gospel, and no deeper cultivation of life than reclaiming the radiance of our being. This means aligning ourselves with our deepest potential for:

- **Rationality** – The ability to reason, think abstractly, and comprehend complex ideas.
- **Creativity** – The capacity to imagine, invent, and produce new works, ideas, or solutions.
- **Moral Agency** – An innate sense of right and wrong that informs ethical choice.
- **Relational Capacity** – The impulse to form deep, meaningful bonds with others.
- **Purposeful Dominion** – The sacred responsibility to steward and cultivate creation.

These traits are not spiritual accessories; they are signatures of divine design. They represent the architecture of our identity, crafted not merely for religious expression but for dominion in every domain of life. Effective marketplace presence begins with a settled knowing - an identity-rooted confidence that precedes performance, anchored in one foundational truth: we were created in the image and likeness of God, endowed with both the capacity and permission to exercise dominion.

This understanding should inform the work of the church, the work of NGOs in developing nations and among the poor in the developed world and what people in those places believe about themselves. Unfortunately, sustaining of the misrepresentation of the poor as culture-bound *homo-sociologicus* to whom the universality of outlook, ambition, and capability cannot be applied persists, often unchallenged and even strengthened by believers and missionaries. In places like Africa and Asia, some claiming to advance the cause of Christ exoticize poverty and dysfunction, which strengthening that some were created less equal than others and belong to species to whom the principle of universality cannot apply.

For believers, the knowing, the identity-rooted confidence that all were created in the image and likeness of God and the capacity and permission to exercise dominion it confers should never be in doubt. Our ability to thrive in the marketplace - to reason, build, lead, design, and multiply - should not be seen as a secular skillset or one belonging to a particular people group, but as sacred inheritance for all. Genesis 1 makes this clear: humanity was created to master life, subdue the earth, be fruitful, multiply, fill, and steward creation. This was presented not as a religious obligation but as an existential reality - a design feature hardwired into the human spirit.

Nowhere in Scripture does God revoke this original intent. In fact, the Great Commission given by Jesus in Matthew 28 harmonizes perfectly with the Genesis mandate. It is not a contradiction but a continuation. The call to make disciples of all nations reasserts the original design: to fill the earth with godly presence, influence, and purpose. From beginning to end, the biblical narrative affirms this truth - we were created with godlike capacity and entrusted with godlike responsibility.

Why, then, does this settled knowing seem absent in so many believers today?

One major culprit is religious miseducation - a well-meaning but damaging approach that treats the Old Testament as irrelevant or intimidating. Many new believers are advised to avoid "the begats," skip the complexity of the Hebrew Scriptures, and focus solely on the Gospels. In doing so, we sever the roots of our identity. The Creation story is not background noise; it is foundational to everything Jesus came to redeem. Without understanding what was first created, we cannot fully grasp the meaning of redemption.

How can we appreciate the cost of The Cross if we do not understand the value of the image it came to restore? Paul emphasizes this point in his second letter to Timothy:

> "But you must continue in the things which you have learned and been assured of, knowing from whom you have learned them, and that from childhood you have known the Holy Scriptures, which are able to make you wise for salvation through faith which is in Christ Jesus." (2 Timothy 3:14-15)

When Paul refers to "the Holy Scriptures," he means texts Timothy was raised on:

- **The Torah (Pentateuch)** – Genesis through Deuteronomy.
- **The Prophets (Nevi'im)** – Historical and prophetic writings.
- **The Writings (Ketuvim)** – Wisdom literature and poetic books like Psalms and Proverbs.

These are not obsolete relics - they are foundational texts. They reveal God's character, the narrative arc of creation and covenant, and the prophetic foreshadowing of the Messiah. They also provide a theological architecture for presence, reminding us that faith was never meant to be detached from life, leadership, work, or witness.

Genesis 3:15, Isaiah 7:14, Isaiah 9:6–7, Isaiah 53, Psalm 22, and Psalm 16 all prophetically point to Jesus. And along the way, we meet Abraham, the father of faith, whose righteousness came not through performance but belief (Genesis 15:6). The Old Testament offers a divine blueprint revealing not only who God is but also who we are and what we are meant to do in the world.

The Old Testament, especially the Creation account, removes any excuse to show up in the world lacking power, purpose, or permission. It affirms that our capacity to lead, build, innovate, and govern is not a secular ambition but a sacred calling.

True presence - decisive, weighty, effectual - flows from this truth. It begins not with religious effort but with settled knowing, awakening to the divine imprint within us and living from it.

The question, then, is not whether believers can show up with presence in the marketplace. The question is: Do we truly know who we are? Tragically, we have allowed the structures of belonging we've created to define our identity and determine our capacity. The research for my doctoral studies in Transformational Leadership, which focused on Africa, found that we have made human systems of belonging - tribes, chiefdoms, and similar structures - *sacred*, even when they obstruct the optimizing effects of faith, education, and innovation.

If the enlightenment of native populations through education and faith contributed to the success of the decolonization process, it has not, in most cases, been extended to the redefinition of the very structures of belonging - structures that continue to shape identity and limit power. Psychologist T. Len Holdstock, in *Re-Examining Psy-*

chology: Critical Perspectives and African Perspectives (2000), refers to these as "the social structures... that prevent the fulfilment of human needs."

Breaking the surly bonds of these systems - which are not unique to the so-called "Third World," but can also be found in what are often called "communities of need" - is no easy task. These include inner-city neighborhoods plagued by generational poverty, indigenous populations marginalized by colonial legacies, and rural enclaves overlooked by national development agendas. In such contexts, inherited structures of identity and belonging - whether tribal, ethnic, denominational, or familial - often provide the only source of affirmation and community. But they can also restrict mobility, reinforce harmful hierarchies, and suppress the very transformation faith and education were meant to ignite.

To answer that, we must return to the beginning - and explore more deeply the mystery of the **dual accounts of creation.**

The "Dual" Accounts of Creation: A Deeper Understanding

Genesis 1 and Genesis 2 describe two dimensions of humanity: one created, the other formed.

In Genesis 1, God created man in His image and likeness - non-material, unseen, yet entirely real. As John 4:24 tells us, God is Spirit; therefore, the being created in His image must also be spirit. This created being, made in the divine likeness, was called man.

In Genesis 2:7, God then formed man from the dust of the earth. This was not a new creation but the clothing of the created man with physical form. Again, he was called man. Two distinct realities - spirit and flesh - sharing one name. Here lies the mystery of personhood: created man and formed man coexist in divine tension.

Created man, like the God who spoke to Israel from the fire at Horeb, was unseen. Moses warned in Deuteronomy 4:15 - 16, "You saw no form when the Lord spoke to you," thus no carved image - no human likeness - could represent Him. God does not resemble formed man. He is Spirit, and so too are we, first and foremost. This distinction is echoed in Jeremiah 1:5:

"Before I formed you in the womb, I knew you; Before you were born, I sanctified you; I ordained you a prophet to the nations."

Jeremiah existed before he was formed. He was known before he was seen, just as Adam was created in Genesis 1 before being formed in Genesis 2. Why does this matter?

The mandate to be fruitful, multiply, and have dominion was given not to the flesh but to the created man - to the divine essence of our humanity. This essence, known and empowered by God before our formation, was intended to shape how Jeremiah lived - not circumstances, limitations, or others' opinions.

Remarkably, God reminded Jeremiah of his essence at the beginning of his calling - long before Jeremiah faced difficulty and questioned God (Jeremiah 12). Like many today, Jeremiah temporarily forgot his divine origin when confronted with hardship, allowing challenges to drown out his settled knowing of who he was.

Yet no redemptive or existential assignment can be fulfilled without this settled knowing.

From the beginning, God knew His people would face monumental challenges, requiring creativity and courage proportionate to those dangers. As Shimon Peres noted about Israel, creativity had to match not the nation's size but the size of its threats. Such resilience does not arise from national pride or religious tradition but from settled knowing: we were created with the capacity and permission to prevail.

Believers should have a distinct advantage here. We know the story - we are created in God's image and have been equipped with divine assurances of success. Paul reminds Timothy, God has not given us "a spirit of fear, but of power and of love and of a sound mind" (2 Timothy 1:7). Peter confirms God's divine power has provided everything needed for life and godliness (2 Peter 1:3).

Our very presence, even before action, should be formidable. This is why the world instinctively resists a believer aware of their true identity. The spirit of the age senses the weight of this knowing. A believer who walks in the blessed assurance of divine identity carries an intimidating authority. Scripture repeatedly underscores this truth: "Do you not know?"

Paul emphasizes urgently: "Do you not know that the saints will judge the world?... that we will judge angels?" (1 Corinthians 6:2–3).

Jesus confronted a hostile mob, reminding them of their divine essence by quoting their scriptures: "Is it not written, 'You are gods'?" (John 10:34–36). Their fury escalated - not because Jesus claimed divinity - but because He reminded them of theirs. Acknowledging your essence demands action. Ignorance may excuse passivity, but truth demands dominion.

To overcome darkness - our own and the world's - we must reclaim the declaration of Genesis 1: we are beings made in God's image, meant not to shrink but to rule.

Much of this truth has been obscured by specialized religious language, creating insider clubs - like Saul's army, trained yet paralyzed before Goliath. But David, unschooled in this system, confronted Goliath with a dangerous clarity: he knew who he was.

Sometimes, contemporary examples clarify this truth. My family's faith foundation was laid by spiritual giants, notably my maternal grandmother, who, despite no formal theological training, knew her essence. When a lion emerged from the bush as she walked home, carrying my mother on her back, she simply declared: "Thank you Lord for sending me a protector," and continued on. The lion followed her to the edge of the courtyard, then retreated.

I am alive today because of a woman who carried her divine essence into every circumstance.

When we embrace our essence, Ecclesiastes 9:11 makes sense. God doesn't level the playing field with equal gifts but provides everyone equal capacity and permission to thrive:

> "I returned and saw under the sun that -
> The race is not to the swift,
> Nor the battle to the strong,
> Nor bread to the wise,
> Nor riches to men of understanding,
> Nor favor to men of skill;
> But time and chance happen to them all."

Larry Bird's improbable dominance in basketball - widely considered a Black man's game - Neil deGrasse Tyson's towering Brown presence in the world of astrophysics, blind Stevie Wonder's musical genius, and the combined 28-year presence of two of the NBA's most unlikely stars - the 5'6" Spud Webb and the 5'3" Muggsy Bogues - all testify to the power of essence over external limitations.

Their success began with settled knowing - the essence within greater than their external frames. They brought this essence into every competitive arena, demonstrating that when essence is missing from presence, we are effectively absent.

Just as David, barely fifteen, faced Goliath not with physical might but with his internal divine image, the true battle always starts before the battlefield. Dominion arises from settled identity.

Reclaiming the divinity of our humanity enables us to step into life's arenas with courage and confidence. It empowers us to make the weight so we can walk in our dominion. The message of our presence to those who are broken, to those who are suffering, is this: you can, too. And the identity that must take precedence is not tribal. Not cultural. Not even religious. It is the true and powerful identity that God graciously shared with us at creation:

Imago Dei. The image of God.

A Leader and kingdom citizen blueprint of Presence

Living from the Weight of Who You Are

1. Know Before You Do

Leadership and redemptive kingdom citizenship begins not with action but with **being**. God revealed Himself to Moses not by task, but by identity: *"I Am who I Am."* Cultivate **settled knowing** - a quiet, unshakable awareness of who you are first instead of trying to define yourself just by productivity, platform, or performance.

◈ *"Am I leading from essence, or just from effort?"*

Leadership Practice: Start every major decision, meeting, or initiative with a return to identity.

2. Carry Weight, Not Just Words

Presence is not volume. It's **weight**.
Like the Hebrew word *kabowd* – glory - it is substance, not spectacle. You don't need to announce who you are when you've embodied who you are.

Presence that radiates redemptive purpose and authority flows from alignment with your design.

◈ *"Do I carry the same weight in silence as I do when I speak?"*

Leadership Practice: Focus less on sounding impressive and more on being internally aligned.

---

3. Be Radiant, Not Reactionary

Presence changes environments without effort. Like uranium on a sidewalk, it doesn't need motion to make impact - it simply **is**.

Marie Curie didn't just contribute knowledge - she *embodied* radiance. Leaders, too, must bring a **transformative presence** into toxic, broken, or resistant systems - not with force, but with essence.

◈ *"Am I leaking energy or radiating presence?"*

Leadership Practice: Audit the atmospheres you enter. Do they shift when you arrive?

---

4. Root Authority in Identity, Not Role

Nehemiah wasn't a governor when he rebuilt the wall. David wasn't a soldier when he faced Goliath. Your **essence precedes your title.**

Leadership grounded in divine image carries authority even before it's recognized by others.

◈ *"Am I leading now, even without the label?"*

Leadership Practice: Don't wait for permission to show up. Practice proactive presence even before promotion.

---

5. Recover the Ground of Being

The world remembers what you do, but Heaven recognizes who you are.

Before re-entering public arenas - education, government, media, marketplace - the Church must **reclaim its lost radiance**. That means recovering a theology of identity. Not religious performance, but a return to the **Imago Dei.**

◈ *"Have I cultivated a theology of presence, or am I relying on personality?"*

Leadership Practice: Root your influence in theology, not charisma.

---

6. Reject the Tyranny of Religious Miseducation

Leaders must overcome teachings that disconnect identity from action.

The Genesis mandate wasn't revoked - it was reaffirmed in the Great Commission.

To lead well, we must reconnect the **Old Testament origins** with **New Testament commissioning**.

◈ *"Do I see my dominion as secular ambition or sacred assignment?"*

Leadership Practice: Integrate both covenants into your leadership framework.

---

7. Be More Than Visible - Be Formidable

Presence is not about being seen. It's about being **unignorable**.

A believer aware of their divine essence intimidates systems of darkness - not because they're loud, but because they're *anchored*.

As the Spirit-filled grandmother facing a lion declared: *"I know who I am."*

So must you.

◈ *"Does my essence make systems nervous?"*

Leadership Practice: Cultivate presence through deep identity work.

❊ Final Charge:

When essence is missing from presence, you are effectively absent.

Let your leadership begin in the Spirit, be grounded in truth, and radiate from settled identity. Stand up. Step forward. Speak without needing to shout.

You are not the shadow of someone else's greatness.

You are image-bearing, essence-carrying, glory-reflecting.

Say it with your life: "Hello. I am."

Prevail

Before the battle, before the fall,
Before the curse could touch us all -
He spoke a word, and formed from dust
A being crowned with sacred trust.

Not made to crawl, but to stand,
Not shaped for chains, but to command.
Not to faint, not to strive
To carry life. To be the light

Though fractured now by fear and pain,
The image stamped on us remains.
Beneath the ash, the fire stays -
Divine lava in our clay.

Rise my soul, defy the shame
The lie you're less forever tame
You are the one He says you are
From you the King is never far

By the truth you dare to heed
Prevail today, Go forth, Godspeed.
Not just to boast you bear the seal
You have the power to take the hill

When darkness speaks, stand unafraid.
When giants taunt, be unbetrayed.
The weight you carry is not fake
It is the radiance of His Name.

So build, and bless, and lift the earth.
Remind the dust of what it's worth.
For we were made by Love's detail,
Not for shame - but to prevail.

Dr. Noah Manyika -2025

CHAPTER 2

Relevant Presence: Meaningfully Here.

"You are the salt of the earth."
Matthew 5:13

Stepping Into the Ring

Now that you've weighed in, know this: making the weight class alone won't win the fight. To prevail, you must bring everything - every trait that reflects the divine architecture of your identity. These traits, outlined in Chapter 1- rationality, creativity, moral agency, relational capacity, and purposeful dominion -were not crafted merely for religious expression, but for dominion in every domain of life.

Presence was never meant to be passive. It is the foundation of relevance. And relevance means stepping into the ring - not just with weight, but with **strategy, strength,** and **skill.** It means showing up, not only spiritually grounded, but socially aware, intellectually engaged, and courageously activated.

Let's now explore what it means to be **meaningfully here.**

In the Game

In Chapter 1, we reclaimed the foundational truth that presence precedes performance. We explored the weight of being - the divine essence within us that reflects the image and likeness of God - and how true authority flows not from activity, but from identity. We examined the distinction between the created and the formed man, recognizing that our capacity to prevail in the world originates not in the flesh, but in the spirit-born essence God knew before we were ever seen. From Genesis to Jeremiah, David to Marie Curie, we witnessed how **settled knowing** - the unwavering awareness of who we are - enables us to step into the world with purpose, authority, and radiant presence. Before we speak, build, lead, or serve, our essence must be present. And when it is, everything changes.

Now, we turn our attention to the next frontier and the second principle of the *Prevail* model: Relevance - the practicality and social applicability of presence - the ability to matter in the real world, to bring weight, insight, and transformation to the places where life is actually lived. It is **presence with intelligence** that connects deeply with real needs and that is meaningfully in the game of life.

We were not created solely for moral agency as we narrowly define it. Moral discernment is part of our divine design - but it alone cannot engage the full complexity of real-world issues, nor can it carry the totality of our redemptive mandate. The moralizer who neglects rationality, creativity, relational capacity, and purposeful dominion becomes a caricature of what God created - and ends up enabling the mocking of God instead of advancing His Kingdom.

Often, the super-spirituality of the moralizer is a symptom of fear - a fear of being found wanting in the other divine faculties that define our personhood. But we cannot offer relevant leadership if we believe that what requires study can be resolved by prayer alone, or that what demands diligent, skillful labor can be replaced by spiritual rituals.

Champions in every sport are almost always students of that sport. They don't rely on natural talent alone. They review tape. Boxers spend hours sparring. They rise before dawn to run, to stay in shape, to sharpen not just instinct, but execution.

This is not to devalue prayer or fasting - they are vital. But so too is **non-spiritual preparation:** intellectual curiosity, the courage to engage complex, real-world problems, and the willingness to form meaningful bonds beyond the safety of our fellowships. Relevant presence requires the full expression of the architecture of our identity. We were created to be radiant - fully and meaningfully here. Only then can we become what Jesus called us to be: the salt of the earth.

Relevant presence is not fearful, but **full and fruitful.** It does not hide behind religious jargon, but radiates into difficult spaces with clarity and grace. It does not reduce the gospel to a private moral code for religious people, but reveals it as a transformative truth for all of life. It does not equate righteousness with retreat, avoidance, or spiritual insulation. Instead, it addresses not only the issues the Church is comfortable with, but every issue that pertains to life and godliness - because the Kingdom of God touches every part of life.

Reclaiming the divinity of our humanity means rejecting irrelevance - not by shouting louder or crafting spiritual soundbites that rhyme, but by cultivating a presence that is felt not only in sanctuaries, but in boardrooms, classrooms, studios, city councils, and community centers. A presence that multiplies. A presence that subdues. A presence that brings order, life, and redemption wherever it goes. A presence that understands both the science and mathematics of life - and knows how to leverage them.

This is the gospel - powerful in time, not afraid of the world, but present in it. Not diluted by relevance, but made incarnate through it - empowering us to prevail.

But when we fail to cultivate this kind of presence - when we shrink back from relevance, neglect the gifts woven into our design, or compartmentalize our faith - we often find ourselves frustrated by the brokenness around us. Instead of looking inward, we look upward with accusation. We turn our pain into a prosecution, pointing the finger at God for outcomes that trace back to our own abdication. This is precisely what we witness in the next passage - a moment of bold complaint from a prophet who, though faithful, still wrestled with the apparent injustice of God's judgments.

Case Number J 12:1-14

Jeremiah - the prophet turned prosecutor - could not have had stronger facts to support his case. In the dock: God Himself. Representing himself, Jeremiah was bold, articulate, and forceful in laying out his complaint:

"Righteous are You, O Lord, when I plead with You; Yet let me talk with You about Your judgments. Why does the way of the wicked prosper? Why are those happy who deal so treacherously?" - Jeremiah 12:1.

According to the facts Jeremiah presented, God clearly had a case to answer. While the righteous - like Jeremiah, faithfully carrying out their kingdom assignments - were suffering, God appeared to be showering blessings on the treacherous. Jeremiah's frustration was unmistakable. In the first two verses alone, he uses "You" and "Your" six times, underscoring the accusatory tone:

"You have planted them, yes, they have taken root; They grow, yes, they bear fruit. You are near in their mouth But far from their mind."
- Jeremiah 12:2.

Let's adjourn from the courtroom for a moment and consult 1 John 2:15–17 to help us understand Jeremiah's case:

"Do not love the world or the things in the world. If anyone loves the world, the love of the Father is not in him. For all that is in the world - the lust of the flesh, the lust of the eyes, and the pride of life - is not of the Father but is of the world. And the world is passing away, and the lust of it; but he who does the will of God abides forever."

Jeremiah believed he did not love the world or the things in it, which made his case against God seem strong. As he reminded God:

"But You, O Lord, know me; You have seen me, And You have tested my heart toward You."
- Jeremiah 12:3

There is no reason to doubt that Jeremiah was doing all the right things as he understood them, loving neither the world nor what was in it. But 1 John 2:16 helps clarify what is meant by "the world"- not merely possessions or success, but the motivations behind them:

- **Lust of the flesh**
- **Lust of the eyes**
- **Pride of life**

These are internal drivers. Scripture is addressing what shapes our pursuits, not just what we possess. Believers must be careful not to condemn those living well as worldly, or to assume prosperity is evidence of wickedness.

If accumulating resources had no redemptive value, Proverbs 13:22 would not declare, "A good man leaves an inheritance to his children's children." Nor would Proverbs 29:2 affirm, "When the righteous are in authority, the people rejoice," if righteous leadership in civic life were irrelevant - or if political ambition were inherently worldly.

Likewise, Job 42:12 tells us: "The Lord blessed the latter days of Job more than his beginning..." detailing Job's massive wealth not as a worldly reward but as a divine restoration. God restored

Job's honor, influence, and abundance as evidence that prosperity and godliness are not mutually exclusive.

If material blessing had no redemptive purpose, God would not have enabled Job to impact future generations, even granting his daughters an inheritance among their brothers - a radical act for that time. The same word for "blessed" used here is found in Genesis 1:28, where God blessed mankind and commissioned them to "be fruitful and multiply; fill the earth and subdue it; have dominion..."

Job's story began in righteousness, not just in wealth:

"There was a man in the land of Uz, whose name was Job; and that man was blameless and upright, and one who feared God and shunned evil... so that this man was the greatest of all the people of the East."
- Job 1:1–3.

And what of the Proverbs 31 woman? Scripture highlights her entrepreneurship, investment acumen, and leadership - not her devotional habits. Today, some would wrongly label her as worldly because of her success. But Scripture elevates her as the gold standard of godly womanhood - one who honors God by engaging fully, meaningfully and fruitfully with life.

Religious miseducation has led many to believe that those who prosper without visible piety must be compromised. Like Jeremiah, many Christians today still ask: "Why do the wicked prosper?" and struggle to understand joy and fruitfulness in those outside the faith. This confusion breeds hesitant ambition and prayerlessness. We question whether excellence, success, and godliness can co-exist. As a result, we stop asking God for the very things that would make our presence full and our leadership relevant.

Ecclesiastes 9:16 adds an important insight:

"Wisdom is better than strength. Nevertheless the poor man's wisdom is despised, and his words are not heard."

This sobering truth underscores the reality that even righteous wisdom can be ignored when it lacks the weight of resources or influence. Prosperity, rightly stewarded, can amplify the voice of the godly and give credibility to the wisdom they carry. As such, relevance is not merely about righteousness - it is also about the capacity to be heard.

Let's return to the courtroom. Jeremiah, still in full prosecutorial mode, presents God as the one who planted the wicked - people who speak of God but have no intention of living for Him. Yet they flourish. Their fruitfulness, in Jeremiah's mind, is proof of God's favoritism. He doesn't attribute their success to discipline, creativity, or divine image-bearing, but to God's unfair partiality. As long as God was unfair, Jeremiah would not bring more to his assignment.

Jeremiah doesn't just complain - he offers sentencing guidelines:

> "Pull them out like sheep for the slaughter,
> And prepare them for the day of slaughter.
> How long will the land mourn…?"
> - Jeremiah 12:3 - 4.

God's response is neither defensive nor coddling. It is direct - and personal. He turns the accusatory "You" back on Jeremiah. In verse 5, God uses "you" six times, making it unmistakably clear who the message is for:

"If you have run with the footmen, and they have wearied you, Then how can you contend with horses? And if in the land of peace, in which you trusted, they wearied you, Then how will you do in the floodplain of the Jordan?" - Jeremiah 12:5.

What exactly was God saying? How could Jeremiah contend with horses - when he was not a horse? Surely, nothing created by God was designed to operate outside the boundaries of its nature - its *umwelt*, the specific world defined by the limits of its species' experience. A dog, no matter how smart or well-trained, remains a dog. Its keen sense of smell may seem extraordinary to us, but it is not miraculous in the animal kingdom. It is simply an expression of its dog-ness. What impresses humans is ordinary within the boundaries of its design.

The Exception to the Rule

But God's answer to Jeremiah hints at something radical: humans are the exception to the rule.

God doesn't offer Jeremiah a theological treatise or a defense of His actions. He simply challenges Jeremiah to consider whether the weariness he feels is evidence of misalignment - not with his calling, but with his understanding of what he was made for. Jeremiah, preaching a theology of limitation rather than possibility, needed a reminder of his own origin story.

That foundation had already been laid in Jeremiah 1:5 - long before he ever put God on trial:

"Before I formed you in the womb, I knew you;
Before you were born, I sanctified you;
I ordained you a prophet to the nations."

This was not just a commissioning - it was a revelation of nature. Jeremiah existed before he was formed. He was known before he was seen. He had been entrusted with a divine mandate before he had performed a single task. God had made him for more.

So when Jeremiah began to falter in chapter 12, it wasn't that God was unfair - it was that Jeremiah was forgetting who he was. God had already given him the words, the authority, and the commission in Jeremiah 1:

"Do not say, 'I am a youth,' For you shall go to all to whom I send you, And whatever I command you, you shall speak. Do not be afraid of their faces, For I am with you to deliver you," says the Lord." - Jeremiah 1:7 - 8

"Then the Lord put forth His hand and touched my mouth, and the Lord said to me: "Behold, I have put My words in your mouth. See, I have this day set you over the nations and over the kingdoms, To root out and to pull down, To destroy and to throw down, To build and to plant." - Jeremiah 1:9–10.

Jeremiah didn't need more evidence of God's faithfulness. He needed to stop measuring God's justice by the metrics of others' success and start measuring his own faithfulness by the weight of what he had already received.

So do we.

We provide relevant leadership when we remember that, while human taxonomy places us within the animal kingdom, we are not merely animals. We alone bear the image of God. We alone carry the mandate of dominion. We are the only species with the capacity to transcend our *umwelt*

- to operate beyond the boundaries of our natural environment, to imagine beyond our senses, to bring more to the table than biology alone would predict.

A bird flies because it was designed to. A fish swims because it must. But humans? We are not birds, yet we soar above the clouds in aircraft we've engineered. We are not fish, yet we dive to the depths in vessels we've constructed. We are not stars, yet we orbit them, study them, and place telescopes and stations in space to explore the cosmos.

And most profoundly, we carry life into barren places. Consider our ventures to the moon - a lifeless expanse where humanity planted its flag and walked. That moment was more than a scientific achievement. It was a declaration: we were made to fill the void, to bring breath and presence to what was once empty. This is not arrogance. It is obedience to the original commission:

"Be fruitful and multiply; fill the earth and subdue it; have dominion…" -Genesis 1:28)

From ancient times, even diminutive mahouts have guided massive elephants, and fierce animals have submitted to the commands of their human trainers. Consider the Roma people - often marginalized, stereotyped, and misunderstood - who nonetheless demonstrate this divine principle in circuses around the world. Though society may fail to honor the image of God in them, even the beasts instinctively respond to the authority placed within them by divine design.

But dominion is never meant to be one-dimensional. The authority to tame animals is just one expression of a far deeper reality. That same divine imprint is meant to liberate every area of life - freeing individuals from the limitations of inherited structures of belonging and enabling transformation in their livelihoods, relationships, and communities.

This is the heart of relevant presence: not performance in a single domain, but the full reclamation of identity in every domain. When we live from the truth of who we are, relevance is no longer something we chase - it's something we carry. We do not beg for a seat at the table; we shape the table, elevate the conversation, and transform the space.

This is what it means to be meaningfully here.

Embodied Dominion: The Relevance of Being Here

God's message to Jeremiah underscores something profound: the importance of bringing our full, meaningful, consequential, God-given essence to bear - *here*, in the world, among the systems and structures of life.

Relevant presence means delivering more than our flesh alone can offer. Our formed self - our flesh - shares common ground with the beasts of the field and the birds of the air. Genesis 2:7 and 2:19 confirm that, like them, we were fashioned from the earth. Furthermore, Genesis 1:22 reveals that they were given a mandate parallel to our formed nature:

> "And God blessed them, saying, 'Be fruitful and multiply, and fill the waters in the seas, and let birds multiply on the earth.'"

Neither our flesh nor theirs was created in God's image and likeness, nor entrusted with the dominion or mandate to subdue the earth. As a result, **our formed self - if disconnected from our divine essence - shares their limitations.** This crucial distinction highlights the extraordinary capacity granted uniquely to humanity.

God's message to Jeremiah makes this clear: "You are the exception to the rule. You can bring more to the fight, provided you tap into what is inherently within you - activated by divine design and empowered by your commissioning."

Our impact is meant to be here. Our redemptive purpose is meant to be lived here.

The *Left Behind* series - popular novels about the rapture written by Tim LaHaye and Jerry B. Jenkins between 1995 and 2007 - and the 2014 film adaptation starring Nicolas Cage sparked important conversations about living right before God. They stirred a sense of urgency and reminded many of the importance of spiritual readiness. But they also cultivated, among some, a distracting preoccupation with when the world will end or when Jesus will return. The danger is that by focusing so heavily on escape, we risk undermining our earthly relevance and the revealing of godlikeness on this side of eternity.

We stop cultivating the fullness of our presence in the name of Jesus. We stifle the imagination and curiosity God gave uniquely to humankind as His co-creators. We stop exploring, stop acquiring knowledge that could help us partner with God in advancing His kingdom. Science and innovation begin to appear as threats to faith and righteousness. Meanwhile, we rely on the diligent work of those we label godless to provide our jobs and extend our lives in hospitals, ironically thanking God for outcomes made possible by people and systems we publicly reject. It's a hypocrisy we often fail to see.

But fulfilling humanity's original divine mandate - to fill the earth with God's image, glory, and redemptive influence - is not a distraction from eternity. It is preparation for it.

A singular focus on becoming "heavenly-minded" through strict religious rule-keeping, at the expense of practical, earthly impact, violates the very Scripture it claims to uphold. Likewise, portraying disengagement from systems and marketplaces as spiritual superiority undermines the purpose for which we were created.

In the Sermon on the Mount, Jesus didn't tell His followers to strive to be relevant. He told them they already were:

"You are the salt of the earth." (Matthew 5:13)

But He immediately followed this affirmation with a warning:

"...But if the salt loses its flavor, how shall it be seasoned? It is then good for nothing..."

We were not created to be "good for nothing." Our presence should not be seen - even by the world - as meaningless or weightless. To live as though we carry no redemptive relevance is to betray the very mission Jesus entrusted to us. Being "good for nothing" undermines the Great Commission because it robs the gospel of its incarnational power - power meant to be embodied in real people, in real places, doing real things that reveal the Kingdom of God.

The Great Commission is not a call to escape the world, but to transform it - to go into **all the world** and disciple nations, baptizing them, teaching them, and demonstrating what it means to live under the reign of a good King. That task cannot be accomplished by people whose presence has no weight, whose lives lack evidence of divine design, and whose leadership never leaves the sanctuary.

We were created by God with full relevance: designed to matter, to build, to bless, to lead. Somewhere along the way, we lost sight of that.

But here is the good news: **the divinity of our humanity has never been revoked.** It remains whole- waiting to be reawakened.

And it is not far. It is as near as breath, as close as within. Reach for it - and prevail.

A Leader and kingdom citizen Blueprint of Relevant Presence

Bringing the Fullness of Your Divine Design into the Real World

1. Show Up Where Life Happens

You weren't created to spectate from the sidelines of culture.

Relevance means presence where it matters - in decision-making rooms, classrooms, studios, courtrooms, city councils, and labs. Jesus didn't send salt to the saltshaker - He sent it to the earth.

◈ *"Where is the earth crying out for salt?"*

Leadership Practice: Identify three "non-religious" spaces in your community where your gifts are needed. Step in. Start small. Show up.

---

2. Engage With Intelligence, Not Just Intention

Relevance is not rooted in zeal alone. It requires insight, skill, preparation, and courage.

Prayer is essential - but **prayer without strategy is spiritual escapism**.

Jesus called His disciples to be "wise as serpents and harmless as doves" (Matthew 10:16).

Your competence is Kingdom currency.

◈ *"Where am I underdeveloped in the name of being spiritual?"*

Leadership Practice: Commit to developing your craft. Study, read, research.

3. Bring the Full Architecture of Your Identity

You are more than a moral agent. You are a carrier of divine faculties:

- **Rationality**
- **Creativity**
- **Moral Agency**
- **Relational Capacity**
- **Purposeful Dominion**

Relevant leaders activate the whole of their design.

To neglect any one of these is to diminish your redemptive potential.

◈ *"Which of these five areas have I neglected? What would it look like to fully bring them to your leadership?"*

Leadership Practice: Do an identity audit.

4. Resist the Temptation to Retreat

Jeremiah didn't struggle because he hated the world. He struggled because he didn't understand the fullness of his assignment *in* it.

God didn't call you to escape the world, but to engage it. Get in the game.

Holiness is not retreat. It's redemptive influence.

◈ *"Does being in this world fill me with anxiety? Why"*

Leadership Practice: Identify one arena you've withdrawn from out of fear or fatigue. Re-enter it intentionally - with Presence, not performance.

---

5. Reclaim Excellence Without Guilt

Too many believers confuse spiritual mediocrity with humility.

The Proverbs 31 woman was a leader, investor, and entrepreneur.

Job's prosperity was God's restoration.

You are not more righteous because you are struggling - or less godly because you succeed.

◈ *"Is there anything I'm not asking God for because I fear being misunderstood?"*

Leadership Practice: Stop apologizing for ambition and being more suspicious of your own motives than even God is.

---

6. Transcend the Limits of Formed Flesh

You were not just formed - you were created.

Animals operate within instinct; humans carry divine image and intent.

God told Jeremiah to contend with horses - because **we were made to defy natural limitations**.

◈ *"Am I operating from my formed self or my created essence?"*

Leadership Practice: When you hit resistance, pause and ask and recalibrate accordingly.

---

7. Measure Your Leadership by Impact, Not Just Intention

Salt is not praised for being in the shaker. It is praised when it changes what it touches. If we don't bring transformation, we are not evil. Just irrelevant.

And Jesus said irrelevant salt is *"good for nothing."*

◈ *"Am I making things better - or just being busy?"*

Leadership Practice: Evaluate your relevance not by how much you're doing, but by how much is changing.

❀ Final Charge:

You were created to **fill the earth, subdue it, and steward it**.

This is not a secular ambition - it is a sacred commission.

Relevant presence is not a brand. It's not optics. It's not clout.

It is the full weight of your God-given identity, intelligently applied to the real world.

So build. Design. Teach. Lead. Govern. Write. Heal. Innovate. Restore.

And when you enter the room, don't just be there.

Be salt.

"I love to think of nature as an unlimited broadcasting station, through which God speaks to us every hour, if we will only tune in."

George Washington Carver

CHAPTER 3
Exploring Presence:
So Much More

"We shall not cease from exploration
And the end of all our exploring
Will be to arrive where we started
And know the place for the first time."
T.S. Elliot – Four Quartets: Little Gidding

What, Courage Man!

We were never meant to tiptoe through life.
We were made to explore. From Eden to Canaan,
from Abraham's journey to the apostles' witness
"to the ends of the earth," the story of humanity
is a story of sacred exploration - of stepping into
the unknown not recklessly, but reverently. Not
fearfully, but faithfully.

But fear always lurks in the shadows of un-
charted paths. And so do the voices that warn,
"Curiosity killed the cat."

Shakespeare flips that warning on its head in *Much Ado About Nothing*, when Claudio says to Benedick:

"What, courage man! What though care killed a cat? Thou hast mettle enough in thee to kill care."

Here, "care" means more than worry. It stands for the anxiety, fear, and false humility that paralyze the human spirit. Claudio's defiance is more than comic relief - it is a declaration: You have what it takes. You were made for more.

You will hear that refrain again and again in this book: you have the capacity and permission - the *mettle* - to exercise dominion. Not domination, but divinely inspired stewardship. And that requires exploration.

We must not retreat from the unknown simply because it feels foreign. God's promises are rarely tucked within the borders of our comfort zones. They lie in lands we have not yet seen, in knowledge we have not yet acquired, and in cultures and contexts we were not born into. Sir Edmund Hillary, the renowned explorer and first to summit Mt. Everest, once observed:

"It is not the mountain we conquer but ourselves."

In other words, true exploration reshapes the explorer. It is as much an inner transformation as it is outer discovery. Helen Keller, the American author and disability rights advocate, echoed this truth:

"Life is either a daring adventure or nothing."

Religion may balk at the idea of adventure, equating it with self-will. But the truth is - **that's how God set it up.** Our hesitation to venture without a mysterious confirmation we call "calling" reveals our amnesia: we have forgotten how we were made, and the mandate we were given.

That is why T.S. Eliot's words in *Four Quartets: Little Gidding* are especially profound that

"...the end of all our exploring
Will be to arrive where we started
And know the place for the first time."

This book invites us to take that sacred journey - to explore our way back to the beginning. So that if we've forgotten what it means to be made in the image and likeness of God - imbued with divine capacity and granted permission to prevail - we may become reacquainted with it, perhaps even for the first time.

This time, we do not journey alone. We venture with Jesus. We are led by the Holy Spirit, who teaches us all things (John 14:26).

It is no coincidence that the most cerebral scientists - those venturing into the vast unknown of space - often speak in reverent, almost religious terms when confronted with the wonder of creation. Even the most unspiritual cannot help but be moved by the immensity and intricacy of what they see.

Scripture tells us about the faith of Abraham. The fulfillment of God's promise could only happen if Abraham ventured. He had to go to become. The Israelites had to journey to inherit. And you must explore to embody the fullness of what God has placed within you.

So what holds us back? Fear. Plain and powerful. As Franklin D. Roosevelt said during the Great Depression: "The only thing we have to fear is fear itself." Fear distorts reality. It magnifies the threat and minimizes our strength.

"The land through which we have gone as spies is a land that devours its inhabitants... and we were like grasshoppers in our own sight, and so we were in their sight." (Numbers 13:32–33)

When we forget who we are, fear rewrites the story. But exploration is not optional. It is the overflow of identity, and the demand of dominion. You were created to seek, to learn, to discover.

So let Claudio's words be rephrased for our generation:

"What, courage man! What though fear killed the grasshopper? Thou hast mettle enough in thee to kill fear."

The Gift of Being Local in a Global Creation

We were made in the image and likeness of a God who fills both heaven and earth. But unlike God, we are finite. Our presence is limited by time, place, and space. We cannot see everything. Though shaped by the hands of an all-knowing Creator, we are not all-knowing ourselves.

We cannot know it all. But that limitation is not a flaw - it's an invitation.

Into our limited frame, God placed something profoundly divine: the capacity to explore.

Because we are not everywhere, we are meant to *go*. Because we do not know, we are meant to *ask*. Because we cannot see the whole picture, we are meant to *seek*.

From the beginning, humanity was called into discovery. Adam and Eve were not placed in a static garden, but in a dynamic world - rivers flowing outward, animals waiting to be named, resources waiting to be uncovered. The creation mandate - to be fruitful, multiply, fill the earth, and subdue it - was more than a call to stewardship; it was a summons to exploration. To draw out hidden potential. To co-labor with God in bringing order, beauty, and wisdom into every corner of His creation.

Exploring presence - stepping beyond the boundaries of what we currently know or where we've come from, whether across cultures, disciplines, technologies, or even within our own emotional and spiritual lives - honors the One who made a world so rich in meaning that it will take eternity to grasp its fullness. It is also how we participate in our Father's business and live out our redemptive mandate.

Our Father's Business

In Luke 2, we find a striking picture of divine exploration. After the Feast of Passover in Jerusalem, Mary and Joseph begin their return to Nazareth - only to realize their twelve-year-old son is missing. After three days of anxious searching, they find Him not lost, but exactly where He intended to be:

> "...sitting in the midst of the teachers, both listening to them and asking them questions." (v.46)

When His distressed parents confront Him, Jesus responds with words that should reshape how we understand presence, purpose, and the pursuit of knowledge:

> "Why did you seek Me? Did you not know that I must be about My Father's business?" (v.49)

The contrast between Jesus and His parents is profound. For Mary and Joseph, the trip to Jerusalem was a religious tradition - a box to be checked. But for Jesus, it was an opportunity to engage. His presence in the temple was not passive - it was inquisitive, intentional, and formative. He listened, asked, and responded. And the result?

"All who heard Him were astonished at His understanding and answers." (v.47)

This was no incidental moment. It was foundational. At just twelve years old, Jesus was not simply attending a feast - He was preparing to reign. He was exploring to understand. Seeking to be ready. The wisdom He displayed was the seed of the authority He would later walk in.

To reign is not merely to rule - it is to be ready. Ready to interpret times and systems, to discern truth, to meet the moment with insight and authority. Jesus wasn't waiting passively for His hour; He was actively pursuing the knowledge and formation necessary for it.

So must we. Exploration is not a luxury for the curious - it is a necessity for those called to dominion. The creation story spells it out clearly: dominion over the birds of the air and the fish of the sea, over every living thing that moves on the earth - subduing the earth - requires great exploits. That's not a metaphor. It's a mandate. And we cannot fulfill that mandate if we fear the unknown, avoid hard questions, or treat curiosity as a threat to faith.

Yet we often do - because of fear. We fear getting it wrong. We fear rejection. We fear what we might discover - about the world, or about ourselves. Fear convinces us that certainty is safer than discovery. It tells us that staying tethered to tradition, culture, or identity markers is more secure than stepping into the mystery of the unknown.

But fear is not from God. The Spirit invites us into boldness, into love that casts out fear, into peace that steadies us as we go. We cannot overcome fear until we stop being proud of the fears we've allowed to define us.

And fear has a twin: **pride**. Pride often disguises itself in culture. It whispers, "I already know enough." It resists the vulnerability of learning. It turns faith into dogma, wonder into suspicion, and difference into threat. I used to joke with my white friends, "I don't jump out of planes, even with three parachutes strapped on. That's what crazy white people do." I masked fear with cultural pride - until the work I was called to required frequent flying. I had to confront what I once wore like a badge. The same happened when my inability to swim nearly cost my daughter her life. I had to humble myself and learn, as an adult, what I once avoided with bravado.

Fear and pride are hard enough to face on their own. But they are reinforced - and even reward-ed - by religious institutions that fear disruption more than they love truth. Many churches have prioritized conformity over curiosity, preserving tradition at the expense of transformation. But God is not bound by institutional comfort. He in-vites us to follow Him **outside the camp**, beyond the curtain, into the wild places where the Spirit moves freely.

This is especially urgent in how we raise the next generation. We are not called to raise chil-dren who are merely well-behaved in church - but world-shapers, formed by faith and fueled by cu-riosity. The household of faith must become a launchpad, not just a refuge. Our children must learn to ask questions with reverence, challenge assumptions with humility, and carry the weight of their presence into classrooms, laboratories, boardrooms, and beyond.

They were born to reign. And like Jesus, they must be shaped by the wisdom that comes through seeking.

To explore is not to stray from the Father's business - it is to take it seriously. For the business of our Father is the flourishing of His creation, the unveiling of His truth, and the redemption of His world. Let us then raise up a generation of holy explorers - those who will not only inherit the kingdom, but know what to do with it.

To raise explorers is not to raise rebels. It is to raise rulers - those equipped to discern truth, apply wisdom, and act with holy confidence in every sphere of life. This is how we prepare the next generation to reign: not by shielding them from the world, but by anchoring them in identity and releasing them into it with curiosity, courage, and calling.

When presence is paired with pursuit - when being fuels bold exploration -we are not only about our Father's business; we become vessels of His exploits in the world.

"...but the people who know their God shall be strong, and carry out great exploits." (Daniel 11:32b)

Mastery with Meaning

Presence, when animated by the divinity of our humanity, does not shrink from mystery. It does not fear questions or treat science and discovery as forbidden fruit. Rather than clinging to what is already known, it moves with bold curiosity - exploring God's creation, engaging His mysteries, and seeking wisdom as an act of faithfulness.

This is what it means to be about our Father's business: not simply repeating old proclamations, but entering deeply into the workings of the world God made. A faith that disconnects people from the systems of life is not holy - it is harmful. When religious culture discourages inquiry, suppresses understanding, or treats knowledge as a threat, it breeds ignorance that disempowers. There can be no dominion without understanding the design of what we are called to steward. And there can be no understanding without exploration.

Revelation is not locked away behind spiritual rituals. It is the unveiling of truth - wherever and however truth reveals itself. It comes through prayer, yes—but also through research, literature, cross-cultural dialogue, data, experiments, and even stargazing. Revelation is the handprint of God made visible. As David wrote:

"The heavens declare the glory of God;
And the firmament shows His handiwork.
Day unto day utters speech,
And night unto night reveals knowledge."
(Psalm 19:1–2)

Creation is not silent. It speaks. And through it, we engage what Paul in Acts 20:27 called *"the whole counsel of God."* Innovation, discovery, and experimentation are not secular indulgences - they are sacred acts. They empower us to move from surviving to cultivating, from observing to understanding, from wandering to prevailing. George Washington Carver, the brilliant American agricultural chemist, agronomist, and experimenter who was born a slave famously said:

"I love to think of nature as an unlimited broadcasting station, through which God speaks to us every hour, if we will only tune in."

This book is not a defense of godless scientism. It is a call to **mastery with meaning** - to reclaim curiosity as a divine trait and to use it in service of God's original design. To prevail is to reclaim our place in the world as those made in God's image - those who have both the capacity and permission to extend His presence into every domain.

To explore faithfully is to pursue God not only in Scripture, but in creation, in culture, in science, and in systems. We explore to see more of Him, to validate the Truth by experiencing His unsearchable depth, and to cultivate Life in a world brimming with redemptive potential.

Exploration is not extracurricular to faith. It is central to our formation as those who reign. This is how we prevail.

Learning from the Heavens

Yuri Alekseyevich Gagarin, a Soviet pilot and cosmonaut, became the first human being to travel into outer space aboard *Vostok 1* on April 12, 1961. Upon returning, he reportedly remarked on the stunning beauty of the Earth as seen from above, urging humanity:

"People, let us preserve and increase this beauty, not destroy it!"

Many who have followed in Gagarin's footsteps - crossing the threshold between Earth and the stars - return changed. They speak of Earth's fragility, its breathtaking unity, and their own transformed sense of purpose. Why? Because exploration expands perspective. It humbles the ego and lifts the soul.

And when we explore with hearts tuned to God, we don't just return smarter - we return **more whole.** This is the essence of exploring presence: to go farther, so we can return wiser. To look outward, so we might live more intentionally inward. To seek knowledge not for dominance, but for stewardship. To wonder, not to escape - but to worship.

NASA astronaut and U.S. Navy test pilot Barry Eugene "Butch" Wilmore recently returned to Earth after nine months aboard the International Space Station (ISS). A mission originally planned for eight days stretched to 280. Wilmore credits his faith - and the ability to join his church's weekend services from space - as what kept him grounded in the midst of such extended isolation.

That human beings can survive 1 hour and 48 minutes in airless space, as Gagarin once did, is remarkable. That they can now endure nearly 300 days is astonishing. But what should no longer be surprising is that such exploration doesn't threaten faith - it can deepen it.

To Butch - and to many others - space exploration doesn't suggest that humanity can do without God. Quite the opposite. It **points to the Creator behind it all.**

NASA astrophysicist Dr. Jennifer Wiseman, a former senior project scientist for the Hubble Space Telescope, has spent her career studying the birthing of stars in distant galaxies. Her work maps the cosmos and helps humanity grasp the scale and intricacy of creation. For her, this is not merely science - it is worship. Her life and work declare a truth many believers overlook: God did not just create this Earth. As Hebrews 11:3 reminds us,

"The worlds [plural] were framed by the word of God."

Dr. Wiseman puts it this way:

"As we explore the universe, we find order, beauty, and complexity that inspire awe."

Her faith and scientific vocation are not in conflict. They are integrated. She approaches the unknown not with suspicion, but with reverence. Her life affirms a powerful truth: Curiosity rooted in trust does not weaken belief. It strengthens it.

To explore is to be fully alive to God's world and deeply loyal to His purposes. It is to respond to the call of mastery with meaning - pursuing wisdom not to control, but to cultivate; not to replace God, but to reveal Him more clearly in the vastness He has made.

Conclusion: Go Far, Come Close.

Exploring presence is not unfaithful - it is profoundly faithful. It is the posture of those who believe God is big enough to be searched out again and again. It is how we grow from glory to glory.

We need pastors who study systems. Scientists who hear the Spirit. Business leaders who ask the questions no one else is asking. Believers who are both grounded and reaching. Because when the people of God stop exploring, they stop leading.

You were not made to know everything. But *you were made to seek.*

Your questions are not a problem - they are *the beginning of wisdom*.

Your limitations are not your enemy - they are *the launchpad for wonder*.

Your curiosity is not a distraction - it is *a divine invitation*.

So go far - so that you can come close. Explore boldly. Discover reverently. Show up present -and searching. The God who made the stars is not afraid of your questions. He is waiting in the mystery, ready to speak to your essence. Go forth. Because there is more.

So much more.

A Leader and kingdom citizen Blueprint of Exploring Presence

Curiosity as a Kingdom Imperative

1. Let Curiosity Be Your Courage

Exploring presence begins where fear ends. The God who formed galaxies is not threatened by your questions. Jesus in the temple wasn't lost -

He was seeking. To explore is not to doubt, but to honor God enough to search for more.

◈ *"What mystery is God inviting me to explore here?"*

Leadership Practice: Begin each new challenge with a question.

---

2. Reject the Illusion of Control

Fear and pride often masquerade as wisdom. Fear says "stay safe." Pride says "you already know." But the Spirit leads only those willing to move. Dominion begins where control ends.

◈ *"What fear or pride has limited my growth here?"*

Leadership Practice: Identify a place in your leadership where comfort has replaced curiosity.

---

3. Make Mastery Meaningful

Dominion is not just about influence - it's about insight. You cannot steward what you do not understand. Curiosity, when guided by faith, leads to revelation with redemptive intent.

◆ **Leadership Practice:** Choose one system you lead in (finance, education, health, culture). Study it deeply as a steward, not a spectator.

---

4. Reframe Revelation

Revelation is not confined to pulpits or prayer closets. It appears in particle accelerators and poetry, constellations and code. Psalm 19 reminds us: the heavens are still speaking. Are we still listening?

◆ *"What does (astronomy, AI, sociology) reveal about the nature of God?"*

Leadership Practice: Pick one unfamiliar field (astronomy, AI, sociology, etc.) and study it as worship.

---

5. Teach Beyond the Textbook

The household of faith should be a launchpad for explorers, not a shelter from truth. Curiosity must be discipled, not dismissed.

◈ *"What's something you're wondering about God or the world?"*

Leadership Practice: In mentoring or discipling others, ask open-ended questions that spark exploration:

---

6. Honor the Whole Counsel of God

Paul didn't just preach doctrine - he understood culture. Daniel mastered Babylonian literature. Jesus asked more questions than He answered. Revelation is layered and expansive.

◈ *"Where might God be speaking that I've ignored?"*

Leadership Practice: Broaden your inputs. Regularly read or engage with something that challenges your assumptions.

---

7. Go Far to Come Close

To explore is to draw nearer to God. Whether you're looking through a telescope or into a spreadsheet, walking the streets of your city or the pages of Scripture - seek Him. He is not hidden from the hungry.

◈ *"What in this expands my understanding of God's world and my role in it?"*

Leadership Practice: Take time each quarter to intentionally explore something new - spiritually, intellectually, culturally.

✥ Final Charge:

Exploring Presence isn't about being impressive - it's about being faithful.

The Kingdom is not advanced by those who cling to certainty, but by those who move with wonder.

The Church doesn't lose influence because we ask too many questions - but because we've stopped asking the ones that matter.

To explore is to worship.
To wonder is to lead well.
To seek is to reign.

So ask. Seek. Knock.
Study. Listen. Stretch. Grow.
Because the God who made the stars is not afraid
of your questions -
He's waiting in the mystery.

"The most extraordinary thing in the
world is an ordinary man
and an ordinary woman and their
ordinary children."

G.K. Chesterton - British author and philosopher

CHAPTER 4

Validating Presence:
You Are More

"...this man, too, is a son of Abraham."
Luke 19:9

You Are Who He Says You Are

In Chapter 1, we talked about **making the weight.** That happens when we know - deeply and securely - that we are who He says we are, and that we are what He made us to be. When we reclaim the divinity of our humanity, we experience true validation. And from that place of wholeness, we are empowered to say to others: "You are who He says you are." In other words, a person walking in the wholeness of being neither takes joy in the feebleness of others (because it presents a "ministry" opportunity) nor feels threatened by their wholeness.

One of the purposes of this work is to challenge us to confront both our absence from the places that shape society, and the motives behind creating **permanent sanctuaries for unhealed wounds** - spaces where pain is preserved rather than transformed, and growth is indefinitely postponed in the name of Jesus.

A gospel that communicates to the world that true healing and wholeness are unattainable is not the gospel. If Jesus were physically walking among us today, it's likely that His presence outside our sanctuaries would empty them - as people rushed to Him to be healed, restored, and to rediscover their dignity outside today's "temple system," finding access to God without religious intermediaries. The cry from the sanctuaries - "The whole world has gone after Him!" (John 12:19) - would not be a cry of praise, just as it wasn't then. It would be panic.

Validating presence rewrites power dynamics and undermines structures built on perpetual need. It tells the lame, "Take up your mat and walk." It tells the marginalized, "Your faith has made you whole." It insists to the outcast, "You are not who they say you are - you are more." It says to the Pharisees concerning Zacchaeus, "... this man, too, is a son of Abraham." (Luke 19:9) It speaks to the Samaritan woman as someone worthy of divine attention - so impacting her life that she ran back to her city and launched a movement.

More about her later.

But first, in the next section, let's briefly examine how the battle lines against validating presence have historically been drawn - and how they continue to be drawn around the faith community today.

The Battlelines

In Exodus 8:28, when Moses demanded that Pharaoh let the children of Israel go, Pharaoh replied, "I will let you go, that you may sacrifice to the Lord your God in the wilderness; only you shall not go very far away." That one phrase - *not very far* - reveals a deeper strategy of spiritual resistance. God wanted the children of Israel to go **all the way** to the Promised Land. Pharaoh just didn't want them to go too far.

Studying both God's original mandate and the life of Christ makes the nature of this battle clear. God empowered us to have dominion - not partial dominion. If partial dominion had been sufficient, we would not have been made in His image and likeness. Consider the helplessness of animals during seasonal shifts - majestic lions unable to roar for lack of strength, brought low by famine, and reduced to prey. That is half-dominion - **subject to the environment**. But humanity was created differently. God made the seasons (Genesis 1:14), but our dominion was never described as seasonal.

Unlike the beasts of the field, we were created to adapt, to strategize, to prevail. Like Isaac, we were made to sow in famine and reap a hundred-fold (Genesis 26:12). Some might argue that "figuring it out" sounds too intellectual, too reliant on reason. But true faith is not blind - it is **rational,** rooted in God's consistent record. Reasoning, when grounded in trust in God's Word, is not anti-faith; it is evidence of it.

So the battle we face is not merely against the devil in the abstract, but against **anything** - and **anyone** - that limits our God-given dominion. While Satan is the ultimate adversary, Scripture records only one direct encounter between him and Jesus: at the beginning of Jesus' public ministry (Matthew 4:1 -11; Mark 1:12 -13; Luke 4:1 - 13). After that, Satan worked through systems and people—most notably, religious opposition.

Jesus didn't just perform miracles; He **declared who He was**. During His conversation with the Samaritan woman at the well in John 4, we read:

"The woman said to Him, 'I know that Messiah is coming (who is called Christ). When He comes, He will tell us all things.' Jesus said to her, 'I who speak to you am He.'" (John 4:25–26)

Validating Presence was Jesus being who He was, **speaking to who she was - deep calling unto deep.** His very presence restored dignity. And that presence posed an existential threat to the religious system. He showed that people mattered more than the systems built around them. He reminded them that the Sabbath was made for man, not man for the Sabbath. Their resistance was rooted in fear of losing control, in pride, and in an unwillingness to relinquish power.

In the end, it was not Rome that initiated Jesus' arrest - it was the religious leaders: chief priests, elders, scribes, and Pharisees (Matthew 26:47–68; Mark 14:43–65; Luke 23:6–25; John 18:28–32; John 19:1–16).

Ironically, it was the religious Jews who turned to the very political system that was oppressing them to do their dirty work and prosecute Jesus. When religion becomes a system of control rather than liberation, it ceases to reflect the heart of God - and God gives us over to our own delusions.

History shows us that when religious systems align with oppressive structures, people turn elsewhere for the validation of their humanity. This created fertile ground for destructive ideologies like Communism. Karl Marx famously declared, "Religion is the opium of the masses," describing it as a tool used to pacify the poor and uphold unjust hierarchies. And he wasn't entirely wrong.

Marx's critique resonated because religion - often stripped of true validating presence - was used to justify inequality and uphold systems that denied some people the dignity of being made in the image of God. When religion supports systems that treat some as fully human and others as lesser, it forfeits its credibility and moral authority. Marx's indictment helped fuel revolutions around the world and gave birth to communist regimes which, though initially built on the promise of justice, became oppressive in their own right.

Communism did not awaken the divinity of humanity - it stoked a fire of anger that built new systems just as broken as the ones they replaced. Jesus came to show a better way: to fight injustice not with vengeance, but with validation.

This is why the message of this book is urgent: the faith community has yet to fully learn this lesson. Failing to validate people creates fertile ground for destructive ideologies and social movements. When leaders prevent followers from discovering truth beyond their sanctioned narratives, it invites rebellion - not just against leaders, but against God.

Today's leaders must realize: they are leading *leaders-in-the-making*. There is no such thing as a permanent follower class. People today are literate, informed, and discerning. They can read Scripture for themselves and see that God did not create an exclusive group and grant them dominion powers while denying them to others. They don't need Karl Marx to help them spot when Scripture is being misused to justify failure. In our time, it's easier than ever to expose the contradictions between declared values and actual practices.

What would our communities look like if our lives, organizations, and influence were rooted in validation, not dependency? What if our ministries didn't preserve problems to justify their existence - but solved them, even if it meant working ourselves out of a job?

The fear shouldn't be that there would be nothing left to do to prove we are merciful, compassionate, or worthy of rapture. Surely, there is a higher calling - a truer calling - than "good works" measured by retention rather than release; by how many stay to maintain enough "tithing units" to cover the church mortgage, rather than how many are sent - healed, whole, and free.

In the research for my Doctor of Transformational Leadership degree, from which the *Prevail* model emerged, I argued that questions about whether NGOs - including faith-based ones - have been a force for good in the developing world are not only justified but necessary. In *The Will to Improve: Governmentality, Development, and the Practice of Politics*, researcher Tania Murray Li observes that organizations working in development in Indonesia (but this could apply globally) rarely stop to examine their actual impact:

"Although improvement seldom lives up to the billing... The endless deferral of the promise of development to the time when the ultimate strategy is devised and implementation perfected does more than enable the development apparatus to sustain itself... It keeps the attention of many critics focused on the deficiencies of such schemes and how to correct them. Meanwhile, *changing the conditions that position some social groups to accumulate while others are impoverished remains firmly off-limits.*" (emphasis mine)

Her critique underscores a haunting reality: some systems are sustained by the very problems they claim to solve. The same can be true in ministry when validation is replaced by perpetual intervention that never restores.

To see what validation looks like when fully embodied, let's now turn to John 4 and witness validating presence in action.

Lessons from the Well

Jesus' interaction with the Samaritan woman is a profound demonstration of validating presence. This was not a random conversation. It was a deliberate, redemptive encounter that stands at the heart of what it means to truly see someone through the lens of heaven.

The woman at the well was, by every social and religious measure of her day, an outsider. She was a Samaritan, part of a people group the Jews regarded as apostates, heretics, and ethnic half-breeds. Samaritans were the descendants of Israelites who had intermarried with surrounding nations after the Assyrian conquest. Their temple worship was considered illegitimate, their scriptures incomplete, and their very identity, to devout Jews, was offensive.

And if being Samaritan wasn't enough to put her on the margins, she was also a woman - in a patriarchal culture where public discourse between unrelated men and women was frowned upon. On top of that, she was a woman with a reputation. Married five times and now living with a man who was not her husband, she was the kind of person people whispered about and avoided. And yet, it is with her that Jesus has this extraordinary exchange:

"When a Samaritan woman came to draw water, Jesus said to her, 'Will you give me a drink?'" (John 4:7)

Immediately, she challenges Him:

"You are a Jew and I am a Samaritan woman. How can you ask me for a drink?" (v.9)

Her response reveals the depth of social and religious division. She defines Him by His ethnicity and herself by her status. But Jesus doesn't engage at that surface level. He replies:

"If you knew the gift of God and who it is that asks you for a drink, you would have asked Him and He would have given you living water." (v.10)

In other words: *"You see me as a Jew and yourself as a Samaritan woman. But if you knew who you really are, and who I really am, this conversation would be different. You are more than you think. And I am here to awaken that truth in you."*

Jesus begins to lead her on a journey of discovery. Not by overwhelming her, but by engaging her in deep, intelligent conversation. She challenges Him. She asks theological questions. She debates the legitimacy of Jewish worship. And Jesus meets her every step of the way, never condescending, never dismissive.

What unfolds is not just an invitation to salvation, but a validation of her capacity.

> "Sir," the woman said, "I can see that you are a prophet. Our ancestors worshiped on this mountain, but you Jews claim that the place where we must worship is in Jerusalem." (v.19–20)

She is not simply curious - she is informed. She has theological convictions. And Jesus honors that. He responds with one of the most important teachings in all of Scripture:

> "Believe me, a time is coming when you will worship the Father neither on this mountain nor in Jerusalem... true worshipers will worship the Father in Spirit and in truth, for they are the kind of worshipers the Father seeks." (v.21–23)

Then comes the pivotal moment we referred to earlier:

> "I know that Messiah" (called Christ) "is coming. When He comes, He will explain everything to us." (v.25)

This is extraordinary. A Samaritan woman, labeled by religion and society, expresses **settled knowledge** of the coming Messiah. She speaks with authority. She speaks with hope. And she includes herself in the promise: "He will explain everything to us."

Jesus replies:

"I, the one speaking to you - I am He." (v.26)

This declarative statement by Jesus was not prideful. It was a statement of both fact and being. At that very moment, the disciples return:

"Just then His disciples returned and were surprised to find Him talking with a woman. But no one asked, 'What do you want?' or 'Why are you talking with her?'" (v.27)

They are shocked - not because they understand what's happening, but because they see her only as a woman, only as a Samaritan. Their perspective is shaped by social norms. Jesus' perspective is shaped by divine truth.

He saw not just her past but her **potential.** He saw not just her status, but her **spirit.** He saw what was formed in Genesis 2 - a living being, yes - but one created in the image and likeness of God, as declared in Genesis 1. He could have chosen to apologize for the treatment of Samaritans by the Jews. He didn't. He could have spent time talking about how wrong it was for women to be treated the way they were. That might have made her feel seen, but not changed.

Jesus was not ignoring injustice. But He knew the most effective way to confront it wasn't simply by lending strength, but by awakening hers. She didn't become His client. She became His co-laborer.

By the end of that conversation, she wasn't just a woman. Not just a Samaritan. She was a powerful child of God.

Validation does more than affirm dignity - it transforms the marginalized into messengers and followers into forerunners. The Samaritan woman didn't just receive a word - she believed it. She understood the purpose of that belief. She became a voice. And she ran.

That is *validating presence.*

The Foundation for Validating Presence

God's validating presence didn't begin in the Gospels. It began in Genesis - woven into the fabric of creation itself. From the beginning, God didn't merely create humanity - He **entrusted** humanity. The declaration in Genesis 1:26 which is referred to repeatedly in this book was not just a creative act; it was a validating act:

"Let us make man in our image, after our likeness. And let them have dominion…"

This was more than divine permission - it was divine validation. God was saying, "You are capable. You are called. You are crowned with the responsibility to rule."

He did not burden humanity with dominion as an impossible weight. He entrusted it as a reflection of identity. What He gave us to steward, He also gave us the capacity and permission to accomplish. That is validation: responsibility matched with identity. Calling grounded in capacity.

But the enemy has always been a master of distortion. Where God validates, Satan undermines.

Where God says, "You are enough because I made you," the enemy whispers, "You are not enough unless you prove it." Where God says, "You have what it takes," the enemy twists it into, "You lack what you need."

The enemy didn't tempt Eve with raw rebellion - he tempted her with doubt about what God had already affirmed. "Did God really say...?" (Genesis 3:1)

But before the fall, before fear, before striving, we see something truly remarkable in Genesis 2:19:

"Out of the ground the Lord God formed every beast of the field and every bird of the air, and brought them to Adam to see what he would call them. And whatever Adam called each living creature, that was its name."

This is staggering. God does not name the creatures Himself. He brings them to Adam. He watches what Adam will call them. And *whatever Adam named them, that was their name.*

This was not just an act of delegation. It was a divine validation of Adam's creative capacity. Adam had no training in semiotics, the study of meaning-making. No exposure to psycholinguistics. No reference points or categories. This wasn't just a naming exercise - it was the first act of creation *ex nihilo* by a human being who was being what God created Him to be – a creator of meaning from nothing.

In my native language, we call a bicycle *bhizautare* - iron horse. The term draws on an existing reference: a horse that transports people, now replaced by a new form made of iron. But Adam had no such parallels. There was no "lion" to compare a tiger to. No "sparrow" to base a robin on. Every name came from within him - not from memory or comparison, but from creative identity.

God didn't correct him. He didn't interrupt. He didn't revise.

> "Whatever Adam called each living creature, that was its name."

Because God was not just testing Adam. He was validating him. Validating the divinity of his humanity. Validating the capacity to **perceive, define, create,** and **govern** the world with wisdom that flowed not from study but from essence. That is the heart of divine validation: not approval after performance, but affirmation of essence before performance.

God's validating presence says, "I made you in My image. You don't have to become something else to be trusted. You already carry what's needed."

Before the fall, before the striving, humanity stood in divine presence -seen, known, and **validated.** That is the presence we are invited to re-discover.

And that is the presence we are called to carry into the lives of others.

Validation in Practice: The JROTC Ethos

Too often, what we call ministry to the poor functions more as maintenance than transformation. When charity is institutionalized without vision for human potential, we reduce people to projects rather than honor them as image-bearers. We comfort pain, but rarely confront potential.

The issue is not that we care for the poor - it's that we often stop at caring. We were never meant to create sanctuaries of sympathy that stall people in survival. We are called to validate, challenge, build, and send.

Fragile leadership builds kingdoms of dependency. Kingdom-minded leadership builds disciples of dominion. This is precisely why the religious establishment feared Jesus. He didn't need an offering to work a miracle. He didn't need a priest's blessing to restore a woman. He bypassed bureaucracy and made wholeness accessible. And that's what made Him dangerous - He revealed what people could become without religious gatekeepers. That same danger still threatens systems that survive by keeping people broken.

Now contrast this with a secular program like the Junior Reserve Officers' Training Corps (JROTC). It's not a charity. It doesn't hand out pity. It calls students into responsibility and future leadership.

"To instill in students... the values of citizenship, service... personal responsibility and a sense of accomplishment."

JROTC doesn't just affirm students' value - it *activates* it. It equips them with tools, gives them a pathway, and rewards those who rise to the challenge. Even students who never enter the military leave with discipline, confidence, communication skills, and a sense of civic worth.

That is what validation looks like in programmatic form: it prepares people for meaningful participation. It honors image-bearers by expecting much from them - because it sees what's already there.

JROTC is a glimpse of what validating presence at scale can look like:

- Seeing capacity, not just need.
- Building systems that produce equity, not just relief.
- Creating movement, not maintenance.

We don't need more programs that pity the poor. We need movements that *activate and honor their potential*. We don't need more spaces that gather people in their wounds. We need leaders who will *call them into their worth*. This is the gospel Jesus preached - *and lived*.

We don't need more programs that pity the poor. We need movements that honor their potential. We don't need more spaces that gather people in their wounds. We need leaders who will call them into their worth. That is the gospel Jesus preached - and lived.

Conclusion: Let the Church Be Dangerous Again

Validation and affirmation may sound alike - but they live in different worlds. Validation speaks to **divine potential.** Affirmation often reinforces comfort. Validation **builds identity.** Affirmation flatters insecurity. Validation **fuels movement.** Affirmation maintains the moment.

Jesus didn't flatter the Samaritan woman, pat Zacchaeus gently on the head, or merely comfort Peter after his failure. He saw their potential and called them into transformation.

That is validating presence: believing that God has placed more within someone than is currently visible - and daring to speak into it, even when it costs comfort.

Validating presence doesn't just congratulate survival - it equips for thriving.

It doesn't sanctify brokenness as identity - it names it as a stage, not a destination.

It doesn't keep people dependent on compassion - it launches them in confidence.

It balances comfort with calling, grace with growth, strength with expectation, and significance with responsibility. It **names** potential. It **invites** contribution. It *expects* greatness - because it sees the image of God.

Jesus never lowered the standard to make people feel better. He raised people to meet their design. Real discipleship isn't soft - it's sacred and strong. It pulls people out of smallness and into strength.

Performative affirmation produces fragile followers and exhausted leaders. The ones being served never step into strength. The ones leading burn out solving problems they were never meant to carry alone.

But validating presence? It multiplies. It releases. It raises up leaders, builders, thinkers, parents, entrepreneurs, and world-changers. It looks like Jesus.

"Neither do I condemn you... now go and leave your life of sin." (John 8:11)

The Church isn't failing for lack of compassion - it's failing for lack of validation. We weren't called to manage brokenness. We were called to awaken image-bearers. Validation threatens systems built on dependency. It disrupts fragile leadership. It releases movements.

Jesus didn't die for more programs. He died to restore purpose.

He didn't come to pat people on the head - He came to *call them by name,* and send them into the world as redeemed forces of restoration.

Let the Church be dangerous again. Not because it controls people, but because it validates them. Not because it holds all the answers, but because it unleashes the ones God already placed within His people. C.S. Lewis in *The Weight of Glory* writes:

"There are no ordinary people. You have never talked to a mere mortal. Nations, cultures, arts, civilization - these are mortal, and their life is to ours as the life of a gnat. But it is immortals whom we joke with, work with, marry, snub, and exploit - immortal horrors or everlasting splendors."

Let our presence **awaken the immortal presence.** Let our leadership release leadership. Let us, like Jesus, **see** the Samaritan - and **send** the evangelist.

What if the Church became a place where:

- The poor weren't pitied, but prepared?
- The broken weren't managed, but made whole?
- The marginalized weren't merely consoled, but commissioned?
- The next generation wasn't entertained, but entrusted?

What if we stopped measuring success by how many came to us - and started measuring it by how many were sent from us, carrying the weight of God's glory, the fullness of their potential, and the confidence that comes from being seen?

Validation is not just a leadership style. It is the heartbeat of heaven. It began in Eden. It was embodied by Christ. Let us leave behind safe, fragile, predictable religion and step boldly into our original design:

Be fruitful. Multiply. Fill the earth. Subdue it.

A Leader and kingdom citizen Blueprint for Validating Presence

The Core Premise *Validating Presence doesn't merely affirm. It activates. It names divine capacity and awakens dormant potential. To validate is to do what Christ did: to see beyond surface stories, social labels, and survival identities—and call forth the image of God within.*

1. Lead to Liberate, Not to Maintain Validation disrupts maintenance-based leadership. Jesus didn't preserve dysfunction to keep His ministry funded; He healed and sent people. Leaders must ask:

◆ *"Are my systems built to retain followers or to release leaders?"*

Leadership Practice: Redesign one core program with a release mindset -shift metrics from attendance to transformation. Replace dependency loops with equipping pipelines.

2. Challenge Fragility with Identity When pain becomes identity, people grow addicted to pity. Jesus didn't leave people in that state. He validated their essence, then called them to rise.

◈ *"Take up your mat and walk" is not a suggestion. It is a validation of capacity.*

Leadership Practice: Speak directly to identity, not just circumstance. Say, *"You are not broken. You are becoming."*

---

3. Break Systems Built on Perpetual Need Validation makes some religious models obsolete. It demands that we evolve or dissolve any ministry that thrives on people staying small.

◈ *"What part of our mission relies on managing brokenness rather than ending it?"*

Leadership Practice: Audit your language, outcomes, and messaging. Remove any message that implies people need your help more than they need God's empowerment.

4. Build Movements, Not Just Ministries Ministries gather. Movements scatter. Jesus did not plant churches; He released ambassadors.

◈ *"Who are we sending? Who are we empowering to become the answer?"*

Leadership Practice: Commission three people this month. Publicly affirm what you see in them. Give them space, voice, and responsibility.

---

5. Restore Naming Power Adam named creation. Jesus renamed Simon. Validation gives people back the authority to define their story.

◈ *"Who in your world needs to be reminded that they can name their future?"*

Leadership Practice: Ask people, *"What do you believe God calls you?"* Then reflect back the names you discern God speaking over them.

6. Disrupt with Dignity Jesus broke rules to re-store worth - touching lepers, defending women, healing on the Sabbath. He wasn't reckless. He was redemptive.

◈ *Validation will make you dangerous to fragile systems.*

Leadership Practice: Identify one social, institutional, or systemic norm your leadership must disrupt. Do it with clarity and courage.

---

7. Train for Release, Not Retention JROTC doesn't baby students. It validates potential through discipline, vision, and standards. The Church must do the same.

◈ *"Are we preparing the poor, or just pitying them?"*

Leadership Practice: Implement high-expectation discipleship. Combine compassion with challenge. Call people up, not just in.

Final Charge To validate is to do more than affirm dignity - it is to **equip identity**. The Church doesn't need more platforms to attract crowds. It needs people, **released in the full weight of their design**, to transform cities, schools, businesses, and systems.

You are not here to manage brokenness. You are here to awaken image-bearers.

So name them. Challenge them. Release them.

That is validating presence. That is Kingdom leadership. That is how we PREVAIL.

"...as compassionate people, we have been evaluating our charity by the rewards we receive through service, rather than the benefits received by the served. We have failed to adequately calculate the effects of our service on the lives of those reduced to objects of our pity and patronage."

Robert D. Lupton - Toxic Charity

CHAPTER 5

Alleviating Presence: When Mercy Meets Identity

Help becomes harm when it
preserves survival but paralyzes purpose.
Alleviation must always be a
bridge - not a bed.

The Full Measure of Mercy

Validating Presence and *Alleviating Presence* are not rivals; they are companions in redemptive leadership and kingdom citizenship. Validation is not a substitute for alleviation - it is the foundation for doing it rightly. Alleviating Presence demands more than compassion; it requires courage, discernment, and a radical commitment to restoration. We alleviate best when we recognize worth first. English author, philosopher, and Christian apologist G.K. Chesterton couldn't have expressed the idea of human worth more powerfully:

"The most extraordinary thing in the world is an ordinary man and an ordinary woman and their ordinary children."

When we see people as image-bearers, not just problems to solve, we are better prepared to serve them with honor and intentionality.

The fact that Jesus didn't establish hospitals for lepers but touched and healed them does not mean medical care or charitable institutions are bad. Rather, He was challenging us never to settle for mercy responses that fall below the full measure of what God has empowered us to do. In other words, we don't withhold healing so we can justify building hospitals. We don't stop creating jobs or finding solutions for housing so we can continue feeding the homeless and unemployed indefinitely.

Matthew 9 also demonstrates how deeply connected alleviation and validation truly are. When the paralytic was brought to Jesus, He addressed a need no one else saw. The man's problem wasn't only physical - he was carrying the burden of guilt and sin. So Jesus said in verse 2:

"Son, be of good cheer; your sins are forgiven you."

The burden of sin may have been lifted, but Jesus didn't say, "I've done enough good," when it was clearly within His power to do more. So we read in verses 6-7:

"Then He said to the paralytic, 'Arise, take up your bed, and go to your house.' And he arose and departed to his house."

Jesus did both so that the scribes who had murmured, "This man blasphemes!" would know that He had authority to both heal and forgive sins.

The challenge this chapter presents is this: What has God empowered us to do beyond the non-threatening religious rituals that may reduce pain - but leave lives unchanged?

The Other Side of the Street

In Luke 10, a man traveling from Jerusalem to Jericho is attacked, beaten, and left half-dead by robbers. Two religious figures - a priest and a Levite - see him, yet pass by. But it is the Samaritan, culturally marginalized and religiously suspect, who stops to help.

When Jesus asks the lawyer to whom He is telling this story, "Which of these three was a neighbor to him who fell among the thieves?" the man replies, "He who showed mercy on him." (v.37) Jesus responds not with a vague call to kindness, but with a summons to redemptive, intentional mercy: "Go and do likewise." (v.37)

What did the Good Samaritan do that Jesus wants the lawyer - and us - to emulate? It's clear:

"He went to him and bandaged his wounds, pouring on oil and wine; and he set him on his own animal, brought him to an inn, and took care of him." (v.34)

He didn't stop at noticing. He touched the wounds. He applied healing agents. He inconvenienced himself - giving up his own ride to carry the man. He didn't merely relocate the man or outsource his care. Scripture continues:

"On the next day, when he departed, he took out two denarii, gave them to the innkeeper, and said to him, 'Take care of him; and whatever more you spend, when I come again, I will repay you.'" (v.35)

He didn't offload responsibility - he extended it. He resourced the next stage of healing and pledged to return. By involving others in the mission to restore what had been broken, he created a network of care.

Had he simply dropped the man off without dressing his wounds or returning, we could still say he helped. But he wasn't merely helping. This was alleviating presence at its best - it began with recognizing worth and with the understanding that help alone does not restore.

The challenge of this parable is that in order to be like the Good Samaritan, we must first choose not to be like the priest or the Levite. We must unlearn the instincts that lead us to respond to suffering with avoidance or indifference.

It's easy to say they didn't care - and that may be partly true. But when we critique the faith community's failure to fully engage the marketplace or meet tangible needs, we must also acknowledge other factors - like **religious busyness.**

Many in ministry experience burnout not because they lack compassion, but because they are consumed with sustaining the institutional machinery of religion - responding to the constant demands of fragile, sometimes entitled members who expect attention simply because they are "card-carrying" members of a religious organization.

Jesus paints a picture of two religious figures too busy to care:

> "Now by chance a certain priest came down that road. And when he saw him, he passed by on the other side. Likewise a Levite, when he arrived at the place, came and looked, and passed by on the other side." (vv.31–32)

The "other side" they chose was the road to their all-consuming religious obligations. Religion keeps us busy. And sometimes, it doesn't even need to blind us - we still see the suffering. The priest saw the man. The Levite came closer and looked. But both kept walking.

We, the people of faith, are often the greatest sustainers of the sacred–secular divide - and, paradoxically, the reason the world accuses us of not caring. We've never fully resolved which side of the street we're supposed to walk on. We profess compassion, yet often outsource care. We preach healing, yet feel inadequate to respond to real wounds. We leave the bleeding to the medics and the broken to the social workers - then show up to preside at the funerals of those who succumbed to injuries we passed by.

Why? Because we've confused our identity as Kingdom citizens with a job description narrowly defined by pulpit and pew. We've limited our spiritual authority to spiritual spaces. But the man left for dead was not lying on the synagogue floor - he was on the roadside. And that's where Kingdom presence is most needed.

When we divorce ministry from marketplace, and compassion from competence, we become observers of suffering instead of agents of restoration. And in doing so, we legitimize the world's suspicion that our gospel is all talk and no touch.

The Good Samaritan, by contrast, is not too busy to care. Clearly, he had somewhere to be - he tells the innkeeper he will return. His journey is not abandoned; it is interrupted by mercy.

So when Jesus says, "Do likewise," He isn't asking us to abandon our journeys entirely. He's not calling everyone to leave their careers and become Mother Teresa - unless God tells you to.

He's saying something both simple and profound: **You can walk and chew gum at the same time.** You can keep moving toward your God-given goals and still stop to bind up the wounded along the way. You can be productive without becoming indifferent. You can fulfill your assignment without ignoring the suffering around you.

Never become too busy to care - even in the name of God.

Street Corners and Scammers: What Now?

How do we respond to real need in a world full of imitation and exploitation? The rise of scammers - posing as desperate parents, miming violin performances while holding signs for rent or medical expenses - exposes a modern dilemma: Who is truly in need? And what does redemptive help really look like?

In the 1990s, the WWJD movement - *What Would Jesus Do?*- reached its zenith. It reminded believers to model their actions on the love of Christ. But like many well-meaning slogans, it eventually became unmoored from its theological foundation. As the height of its popularity, some in the faith community began to grow tired of being guilted into expressions of mercy that didn't reflect God's wisdom - efforts that didn't point the way to God, validate truth, or cultivate life.

Once appropriated by the world, *WWJD* was no longer a sacred prompt for Christlike action - it became a secular accusation against the Church. It was used by those who never truly sought to understand what Christ did, but were quick to point out what Christians weren't doing.

For *Alleviating Presence* to be truly redemptive, it can't be based on speculation about what Jesus **might** do. It must be rooted in what He **actually did** - in creation, in His earthly ministry, and most fully on the Cross.

In Luke 10, "Go and do likewise" pointed to the Good Samaritan as the model. But in many of Jesus' parables, certain characters are clear stand-ins for Himself. In Matthew 13, He is the Sower. In the parable of the Tares, He explicitly identifies Himself as the farmer:

"He who sows the good seed is the Son of Man." (Matthew 13:37)

So it is in Luke 10. The contrast between the Samaritan and the religious leaders mirrors Jesus' own ministry. Like the Samaritan, Jesus was marginalized and religiously suspect - a Nazarene from whom not much was expected. As Nathanael said:

"Can anything good come out of Nazareth?" (John 1:46)

Jesus embodied Alleviating Presence. His mercy had weight. His actions were redemptive, not performative. And sometimes, His response was silence.

In Matthew 13:58, after being rejected in Nazareth, we read:

"Now He did not do many mighty works there because of their unbelief."

It wasn't because they didn't need miracles. They did. But need alone does not obligate God's intervention. Jesus never responded to need alone. He responded to faith, identity, and alignment with purpose.

This is why WWJD can never be answered by sentiment. It must be shaped by WDJD - **What**

Did Jesus Do? And the answer is: He acted with clarity, purpose, and Spirit-led discernment.

We don't correct our failure to love by adopting the world's definition of love. We don't prove we care by conforming to guilt-driven demands that lack divine alignment. Jesus, who drove out the money changers and withheld miracles in the face of unbelief, is unlikely to reward systems of exploitation simply to look good.

Entire generations and communities have been raised on a culture of dependency, even reinforced by religious institutions. But the existence of these patterns is not an excuse to perpetuate them. We are not called to soothe dysfunction - we are called to redeem it.

In recent years, many in the faith community have recognized that reliance on external funding for faith-based work comes with strings attached. The same pipelines of provision can be shut off by the very systems that supplied them. And when mercy becomes unsustainable - when budget cuts come as needs grow - we're forced to confront a hard truth: **The wrong kind of alleviation deepens the problem.**

Alleviation rooted in distorted, sentimental versions of love preserves pain instead of transforming it. It reduces people to containers of endless compassion, where no care is ever enough because no transformation ever occurs. Help becomes harm when it preserves survival but paralyzes purpose. Alleviation must always be a bridge - not a bed.

Whether we are brave enough to pivot or not, the day of reckoning is already here. We must awaken the divinity of those we serve - and equip them to become agents of their own rescue. This means embracing not only Acts 20:35:

"You must support the weak,"

but also 2 Thessalonians 3:10:

"If anyone will not work, neither shall he eat."

One does not cancel the other. The latter is not cruelty - it is clarity. It's the gospel. We support the weak not so they remain weak, but so they can remember their strength. True giving is not rooted in pity - it's rooted in partnership.

We are not called to soothe suffering indiscriminately, but to align with God's redemptive purpose in each person's life.

We should no longer feel guilty for withholding resources from able-bodied people begging beside "Now Hiring" signs. Love for the sake of love may sound poetic, but Jesus' love had a purpose. His compassion was never about sustaining victimhood. It was about restoring destiny.

The Good Samaritan - if he is indeed a picture of Jesus - was fully invested in the outcome. He extended a network of care, resourced the process of healing, and pledged to return. That is what Christlike alleviation looks like.

A day is coming - and may already be here - when, like Peter and John in Acts 3, we won't have what is being asked for, but we will have what is needed. When they encountered a lame man begging for alms, Peter said:

"Silver and gold I do not have, but what I have I give you: In the name of Jesus Christ of Nazareth, rise up and walk." (v.6)

Peter didn't give what was expected. He gave what was transformational. Alleviating Presence dares to pause, listen, and ask: "What do I have that would truly help?" Reflexive giving may soothe guilt, but redemptive giving seeks restoration. It doesn't just relieve suffering - it reawakens destiny. It doesn't respond to pressure - it listens for heaven.

The Church That Has What the World Needs

Imagine if the Church today operated with the same clarity and intentionality as Peter in Acts 3. What if we stopped asking, "What do they want?" and instead asked, "What have we been given?"

If the Church had consistently helped those it serves reclaim the divinity of their humanity - if it had seen them as image-bearers, not just attenders or recipients - then sitting in every pew would be leaders, innovators, builders, and healers. If everything we did helped people discover that the deepest truth about them isn't where they were born, to whom they were born, their race, gender, or struggles - but an eternal essence crafted before time, in the very image of God - it would awaken the capacity God embedded in each of us at creation.

And when that capacity is awakened, the Church becomes a powerful force and storehouse of spiritual, intellectual, and practical resources - ready not just to serve the world, but to transform it.

What would we discover?

We would find **teachers,** not just congregants - people ready to disciple the next generation. **Entrepreneurs,** not just tithers - men and women who could create businesses that bring dignity through employment. **Counselors, artists, coders, farmers,** and **visionaries** - waiting to be seen, equipped, and released. We would find stories of resilience ready to become testimonies. Ideas that solve problems. Dreams that build futures. Strategies that heal systems - not because we imported answers, but because we ignited what God had already placed within.

If the Church had truly explored the divine potential of its people - not just as members but as ministers, not just as wounded but as warriors - we would already be equipped to respond to every form of suffering.

We would no longer feel helpless in the face of poverty, addiction, or violence. We would see that we have voices that can mentor, minds that can strategize, and hands that can build. We wouldn't just give handouts - we would hand over **keys**. The problem is not that the Church lacks resources. It's that we haven't recognized the treasure within our reach.

Peter could say, "What I have, I give you," because he knew what he carried. He had been **present** with Jesus. He was **grounded** in truth. He was **empowered** by the Spirit. He wasn't guessing - he was giving from a place of clarity, radiancy and confidence.

When the Church walks in the fullness of its design - when it validates people into wholeness, equips them with wisdom, explores their God-given potential, and remains rooted in Presence -it will never run out of what the world needs most. We may not always have silver or gold. But we will have vision.

We will have wisdom. We will have presence. We will have people - empowered, anointed, and ready to respond. That will always be enough.

That is what *Alleviating Presence* dares to believe: That when the Church becomes fully itself, the answer is not far away - it is already in the room.

Alleviating Presence and the Redemptive Pattern of Wholeness

In Matthew 14, we see Jesus demonstrate a holistic compassion that did not end with healing the sick. After restoring physical health to a great multitude, He also saw their hunger. When the disciples suggested He send the crowds away to find food, Jesus said:

"You give them something to eat." (v.16)

Healing wasn't enough. Hunger mattered too. Jesus would not separate physical restoration from spiritual and emotional wholeness. He knew that true alleviation doesn't fragment people's needs. He met them *fully* - body, soul, and spirit. Every act of care, every miracle, was a message: this is what it looks like to bring heaven to earth.

We distort the Gospel when we stop at sympathy. Jesus didn't.

And neither can we.

The only way the Church can offer this kind of holistic ministry is by activating the divine agency of its members. It won't be accomplished through programs alone - especially when programs often rise and fall with funding. It will be accomplished through people who, once they reclaim the divinity of their humanity, become radiant gifts from God - capable of transforming lives.

This is why *Alleviating Presence* works hand in hand with *Validating Presence*. It sees not just the suffering, but the strength beneath the surface. It believes that the greatest act of compassion is not only to help the hurting, but to **help them remember who they truly are.**

Few verses are more misunderstood than Matthew 26:11:

"The poor you will always have with you."

Taken out of context, it sounds like resignation. But Jesus wasn't dismissing the poor - He was elevating worship. When the disciples objected to a woman pouring expensive perfume on Him, claiming the money could've gone to the poor, Jesus replied: "You will always have the poor, but you will not always have Me."

What He meant was simple: *"If you see Me rightly, you'll serve them rightly."* His presence fuels our purpose. Without Him, we give things. With Him, we give transformation.

The enduring presence of the poor is not proof that poverty is God's will. It's proof that the Church has not yet fully stepped into its call. If the poor are always with us, it's not because God wills it - but because we have not yet walked in the fullness of the Gospel that dignifies, develops, and delivers.

Alleviating Presence says to every suffering soul: "You are not forgotten. You are not useless. You are important."

True compassion isn't driven by guilt - it's driven by glory, the weight of God's presence. It doesn't sentimentalize suffering. It restores destiny. It doesn't create dependency. It reawakens dominion. True alleviating presence does not stay at the site of suffering. It walks with people toward strength - until they no longer need our presence, only their own."

This is the Church Jesus is building.

A Church that sees clearly.

A Church that walks with power.

A Church that binds wounds and builds lives.

A Church that not only comforts the broken, but sends them out whole.

So they may prevail.

A Leader and kingdom citizen Blueprint for Alleviating Presence

When Mercy Meets Identity

1. Let Validation Precede Alleviation

Compassion without identity creates dependency. Mercy without validation creates fragility. Jesus addressed both the sin and suffering of the paralytic. He forgave *and* healed.

◈ *"Do you see the image of God before you address the pain of man?"*

Leadership Practice: Before you respond to a need, ask: "Have I affirmed this person's worth - or just responded to their wound?"

2. Refuse to Pass on the Other Side

The priest and the Levite saw - but chose safety. The Samaritan saw - and stepped in. Religious busyness and insecurity often keep us from redemptive engagement.

◈ *"Have you mistaken routine ministry for faithful presence?"*

Leadership Practice: Identify one real-world need you've been avoiding. Ask: "What's kept me from responding - and what would faithfulness look like now?"

---

3. Discern Real Need from Exploitation

The presence of pain does not always mean the presence of truth. Jesus withheld miracles in Nazareth - not from spite, but from spiritual clarity.

◈ *"Are you giving from guilt or from discernment?"*

Leadership Practice: Build a Spirit-led decision filter: "Does this act align with God's redemptive purpose - or just my emotional discomfort?"

4. Move from Sympathy to Restoration

Jesus didn't just notice hunger - He fed. He didn't just acknowledge storms - He calmed. Restoration is the target, not emotional release.

◈ *"Are you offering comfort - or catalyzing transformation?"*

Leadership Practice: Evaluate one program or initiative. Ask: "Does this lead people to healing and strength - or does it perpetuate need?"

---

5. Activate the Divine Agency in Others

Alleviating Presence is not heroism - it's empowerment. The greatest mercy is not help, but activation. Your people carry untapped redemptive potential.

◈ *"Are you helping people remember who they are?"*

Leadership Practice: Identify three overlooked or underutilized people. Ask: "What truth can I speak - or opportunity can I create - to awaken their design?"

6. Give What You Carry, Not What Culture Demands

Peter didn't give what was expected - he gave what he carried: power, not coins. Real giving flows from Presence, not pressure.

◈ *"Are you performing to meet expectations - or releasing what the Spirit has placed in you?"*

Leadership Practice: In a moment of need, pause and pray: "Lord, what do I have to give that aligns with Your purpose for this person?"

7. Alleviate With Expectation, Not Pity

The Good Samaritan resourced restoration - not relief. Jesus' compassion always aimed at wholeness. Redemptive giving expects results.

◈ *"Are you expecting transformation - or settling for emotional relief?"*

Leadership Practice: When you give time, resources, or encouragement, ask: "What is the redemptive outcome I'm praying and planning toward?"

✨ Final Charge

Alleviating Presence is not just kindness.
It is Kingdom clarity.

It doesn't stop at feeding the hungry—it prepares them to plant.
It doesn't celebrate wounds—it calls people to rise.
It doesn't soothe brokenness—it restores divine purpose.

We are not called to manage pain.
We are called to awaken presence.
To redeem stories.
To build lives.

Let your leadership bind wounds, break cycles, and unleash destiny.
So that those you serve won't just survive—
They'll prevail.

"The task of leadership is not to put greatness into people, but to elicit it - for the greatness is there already."

John Buchan -Scottish novelist, historian

CHAPTER 6

Inspiring Presence

"They immediately left their nets and followed Him."
Matthew 4:20

Again I say To Be

One of the goals of this book is to help us understand that even what we envy in others, we already have the **capacity to become** because of our essence. We would save ourselves much wasted energy - on envy, comparison, and regret - if we focused on awakening that essence instead of lamenting the personality, athleticism, eloquence, or privileged beginnings we think we were denied by God. The truth is, those attributes are not the source of inspiring presence - essence is.

The biggest challenge in "learning" to be an inspiring person is that we can never reproduce the conditions that revealed the courage, clarity, and moral conviction of the people we admire. Their inspiration wasn't manufactured - it was revealed in response to the unique demands of their moment. We are not called to copy their conditions, but to activate our own calling.

This is why we often struggle to sustain the "inspiration" we feel after Sunday morning messages that call us to be the Davids of our generation. We feel momentarily stirred to slay giants and bring hope to broken places - but by Monday, the marketplace reminds us that **a spark is not a flame.** Reality whispers: *Stay in your lane.* These stories, we are told - by the systems around us or even our inner doubts - are metaphors: spiritually uplifting, but untranslatable to the grit and complexity of daily life. By Thursday, we're limping toward the weekend, waiting for another sermon to *fix* us.

I'll admit: I had my reservations about the title of this chapter. I was concerned it might feed the demand to be inspired when what we need is to be transformed. As Paul writes in 2 Corinthians 5:17:

"Therefore, if anyone is in Christ, he is a new creation; old things have passed away; behold, all things have become new."

Salvation transforms. And that transformation restores the capacity and permission tied to the divinity of our humanity. Whatever we lost in the fall - the capacity to reflect the creative, life-giving nature of God - we regain in the new birth. The call to be an inspiring presence is not a call to exceptional training or elite gifting. It is a call to be who you were created to be.

This only feels like an unrealistic standard if we reject two truths:

1. That we were created in the image and likeness of God.
2. That salvation restores our capacity to live from that image.

I ultimately stayed with the chapter title because I define "inspiring presence" not as the capacity to excite others, but as the capacity to give life. Once life is given, it is not toggled on and off. Life is a state of being, a continuous presence until death. Jesus said in John 10:10:

"I have come that they may have life, and that they may have it more abundantly."

In Mark 11, Jesus curses a fig tree for bearing no fruit. It's a curious act, especially since it wasn't the season for figs. Yet Jesus says: "Let no one eat fruit from you ever again." The next day, the disciples see that the tree has withered. Peter, surprised, points it out. Jesus replies:

> "Have faith in God. For assuredly, I say to you, whoever says to this mountain, 'Be removed and be cast into the sea,' and does not doubt in his heart... he will have whatever he says."
> (Mark 11:22–24)

Jesus is teaching us something extraordinary. The potential to act with authority, to move what seems immovable, is not reserved for an elite few. He uses the word *whoever* - because the potential to give life, to call forth change, to carry inspiring presence, exists in **every person**.

Whoever your heroes are - Winston Churchill, Martin Luther King Jr., Joan of Arc, or the young shepherd David - their enduring inspiration came not from charisma, but from the awakening of their divine design. It was the activation of something already within them.

And that same opportunity is available to you.

The call to be an inspiring presence in your generation begins not with striving, but with obedience to the most fundamental call we all share:

To be.

The Gift of Gifts

When David arrived at the Valley of Elah, 1 Samuel 17 describes the situation he encountered:

"Now the Philistines gathered their armies together to battle... And Saul and the men of Israel... drew up in battle array... The Philistines stood on a mountain on one side, and Israel stood on a mountain on the other side, with a valley between them." (vv.1–3)

Then Goliath appeared - a giant in stature and a "champion" by title (v.4) -issuing his challenge to Israel's army. He repeated it for forty days, morning and evening (v.16), yet not a single man stepped forward. While it may seem like Israel alone was paralyzed by fear, the Philistines may not have been as formidable as they appeared. They seemed to understand that one man - even a giant - does not make an army. After all, they too remained locked in a standoff. If they had been truly confident, they might have attacked. But for a full month and ten days, they didn't.

In all that time, no one in Israel's camp seemed awakened to the divinity of their humanity, nor to the covenant promises God had made to them. Then David arrived.

It's important to pay attention to how he framed his questions - and why they provoked his brother Eliab's anger:

"Then David spoke to the men who stood by him, saying, 'What shall be done for the man who kills this Philistine and takes away the reproach from Israel? For who is this uncircumcised Philistine, that he should defy the armies of the living God?'" (v.26)

David's question wasn't just about the reward - it was a challenge. His very asking exposed the cowardice of those who hadn't acted, even when a reward had been promised. Eliab's angry response reveals this underlying shame:

"Now Eliab his oldest brother heard... and Eliab's anger was aroused... 'Why did you come down here? And with whom have you left those few sheep...? I know your pride and the insolence of your heart.'" (v.28)

But look again: David belonged in that moment - not because of age, training, or profession - but because of something deeper. He didn't speak with the uncertainty of an imposter. Inspiring presence requires overcoming the inner conflict that arises when society reminds us to "stay in our lane" - to stick to the church choir, hand out tracts, pray during disasters, or offer food and comfort - but never challenge systems or face giants.

Yet in Genesis 1–2, God never talks about lanes. He gives the dominion mandate to all, regardless of role or gifting.

So when Eliab accused him, David simply replied: "What have I done now? Is there not a cause?" (v.29)

Not: *"I'm qualified."*
Not: *"I belong here."*
Just: *"Is there not a cause?"*

Saul's army was filled with soldiers. Some were undoubtedly gifted. But neither gifting nor armor produced courage. It was **settled identity** that gave David the boldness to step forward.

To inspire others - to be a giant-slayer - requires confidence in this truth: The dominion mandate was given to all.

Marie Curie stepped into scientific spaces long reserved for men and inspired generations by doing what had once seemed impossible. She didn't demand permission - she acted on what had already been given at creation. That is the gift of inspiring presence: to awaken the unseen capacity in ourselves and others.

In 2021, Elon Musk revealed that he has Asperger's syndrome, part of the autism spectrum. People with such diagnoses are often underestimated - until they reshape the world. Why? Because God did not deny them what He gave us all: **the Gift of Gifts** - His image and likeness, the capacity and permission to prevail.

David didn't debate Eliab. He simply turned away.

> "Then he turned from him toward another and said the same thing; and these people answered him as the first ones did. Now when the words which David spoke were heard, they reported them to Saul; and he sent for him." (vv.30–31)

David understood a vital truth: **Giants don't fall when Davids stay in their lane.** As we discussed in Chapter 1, inspiring action is rooted not in charisma or training, but in a **settled knowing** of who we are. That kind of confidence releases courage. And that courage - grounded in essence, not ego - is what makes inspiring presence possible.

The Generic and the Particular

In 1 Corinthians 12:4–6, Paul explains the diversity of gifts:

"There are differences of ministries, but the same Lord. And there are diversities of activities, but it is the same God who works all in all."

At first glance, this might seem like a call to "stay in our lane." But as Jesus said in John 10:35, "Scripture cannot be broken." There is no contradiction between God assigning distinct roles through spiritual gifts and God granting all humanity, in Genesis, the same foundational mandate: **dominion.** Paul's emphasis on distinctiveness doesn't cancel the universal empowerment bestowed upon us as image-bearers. The particular gifts we carry only find their full power when built upon the generic Gift of Gifts - the Imago Dei.

If, because of the fall, we lost our understanding of the capacitation and permission to reign, the Cross restores it. That's why we sing from our depth:

"I can do all things through Christ who strengthens me." (Philippians 4:13)

Because this Gift of Gifts is given to all, I refer to it as the "generic" gift - not because it is vague, weak, or common, but because it is universal. Let us not confuse "generic" with inferior. This gift is anything but nameless - it is the Imago Dei, the image and likeness of God. And what could carry more weight than that?

The power to inspire is lost when we forget who we represent. Representing the God-brand is not about wearing fish decals or quoting Christian slogans. It's about our presence - our identity - impacting people and transforming the challenges of our times because we have become who we were created to be. We have reclaimed the divinity of our humanity.

David stood in the valley of Elah. Saul's army cowered - not because they lacked weapons, but because they forgot who they were. Inspiring Presence shuts the mouths of mockers. David's confidence was rooted not in armor, but in identity - and that confidence awakened a nation.

The absence of a particular gift cannot excuse us from acting when our being and presence gives us the capacity to intervene. The priest and the Levite in Luke 10 could not plead, "We didn't help because we lack the gift of mercy." God has never accepted identity-based amnesia as a valid excuse.

When Moses objected to his calling, citing his speech limitations, God replied:

"See, I have made you as God to Pharaoh." (Exodus 7:1)

Likewise, we are sent into situations not as spectators, but as representatives of divine authority. We are not gods because of training, but because of *identity*.

Because we bear His image. Because we were made to prevail.

Reconciling Affective-ness and Effectiveness

There is a powerful capacity to respond - not with what our giftedness suggests, but with what the moment demands - when it is the divinity of our humanity that leads. When our actions are not **gift-led** but **being-led,** we show up like Christ: offering that rare balance of *affective* and *effective* leadership that complex moments require. Winston Churchill's leadership during crisis echoed this - embodying a Jesus-model that felt deeply but did not let emotion alone dictate decisions. He acted decisively, yet with compassion. His presence didn't merely stir emotions; it awakened identity.

The moment we are in, demands that we no longer privilege either affective-ness or effective-ness in how we serve communities or disciple nations. The *PREVAIL* model insists on *Inspiring Presence* - which brings truth with grace and grace with truth.

This is the Jesus-model of Inspiring Presence. We see its transformative power most clearly in the life of Peter. Jesus didn't wait for Peter to be perfect. He didn't coddle him when he failed. He kept showing up - breathing life into Peter's fragile identity until Peter didn't just believe - **he knew.**

Inspiring Presence moves people from emotion to essence, from belief to knowing. It calls out the hidden parts of the soul - not with hype, but with holy invitation. It strengthens weak hands. It awakens hungers people didn't know they had. As John Buchan, the Scottish novelist and historian put it: "The task of leadership is not to put greatness into people, but to elicit it – for the greatness is there already."

Peter is often remembered for his failures - his impulsiveness, his denial of Christ - but we forget his faith. If Jesus had only been exciting, Peter would've faded. But Peter became a pillar. We also can't overlook how it all began.

In Matthew 4:18-20, Jesus calls him with simple words:

"Follow Me, and I will make you fishers of men."

And Peter left everything. Why? Not because of persuasion. But because of presence.

Inspiring Presence is not charisma. It is depth awakening depth. It is the call that stirs the divinity lying dormant in our lives while the storms buffet the boats of our circumstances. Inspiring Presence moves people. It changes trajectories. We leave our nets because of it - and follow.

Even as we journey with Him, it empowers us to make destiny-defining decisions. When one disciple asked to first go bury his father, Jesus replied:

"Follow Me, and let the dead bury their own dead." (Matthew 8:22)

Later, when Jesus said, "He who eats My flesh and drinks My blood abides in Me, and I in him..." (John 6:56), many walked away. But Peter stayed. When Jesus asked if the twelve would leave too, Peter responded:

"Lord, to whom shall we go? You have the words of eternal life. Also, we have come to believe and know that You are the Christ." (John 6:68–69)

That's what Inspiring Presence produces: not just belief, but knowing.

True inspiring presence passes the test of betrayal. It survives denial. It walks through desertion and still endures - so that those it once inspired, who may have been pulled away, can find it still standing when they return, ready to complete their journey of formation.

Peter's journey mirrors our own: from enthusiasm to belief, through unbelief, until knowing is forged. And that knowing births a new kind of presence - the Peter of Acts 3, no longer hesitant or ashamed. He speaks to a lame beggar at the temple gate and declares:

> "Silver and gold I do not have, but what I do have I give you: In the name of Jesus Christ of Nazareth, rise up and walk." (v.6)

That's not doubt speaking. That's not excitement. That's **knowing**.

The Upper Room Wasn't Just Exciting

We long for the excitement of Pentecost, but Acts 2 wasn't about energy - it was about transformation. Peter went from denier to declarer:

"Let all the house of Israel know assuredly…"
(Acts 2:36)

He became dangerous - not because he shouted, but because he *knew*. That's what Inspiring Presence does. It fills people with courage and conviction. It doesn't just prepare us for the blessing - it prepares us for the backlash.

Jesus promised that our earthly assignment would come with blessings - "…a hundredfold… with persecutions…" (Mark 10:30). We like the hundredfold blessings. We skip the hardship. But true inspiration prepares us for both. It doesn't just lift us - it fortifies us, even in suffering.

"For when I am weak, then I am strong." (2 Corinthians 12:10)

The Church doesn't need louder voices. It needs deeper presence. Excitement fades. Inspiration *forms*.

May we be formed.
May our presence awaken essence.
May our lives become a holy disruption.

And may the world, encountering us, *know* - not just believe - that Jesus lives.
And we do too.

A Leader and kingdom citizenship Blueprint for Inspiring Presence

When Being Gives Life

1. Live from Essence, Not Envy

Inspiring leaders are not mimics. They do not strive to become someone else - they awaken who they already are. Envy fades when essence is awakened.

◈ *"What do I already carry that I've been ignoring?"*

Leadership Practice: Identify one area where you've compared yourself to others.

2. Move from Spark to Flame

Sunday inspiration without Monday identity will always fizzle. Leaders anchored in essence don't need weekly refueling - they carry fire because they know who they are.

◈ *"How will I carry this flame into the realities of tomorrow?"*

Leadership Practice: After your next spiritual high point, write down the answer to the above question.

---

3. Let Presence Precede Proving

David didn't step forward with qualifications - he stepped forward with identity. Inspiring Presence doesn't announce its resume; it shows up in strength born of knowing.

◈ *"Is this a moment where I'm called to show up in identity rather than expertise?"*

Leadership Practice: The next time you feel unqualified, ask yourself the above question.

4. Don't Stay in Your Lane - Walk in Your Mandate

The world tells you to stay in your lane. God says, "Be fruitful. Multiply. Subdue." Inspiring leaders walk where God sends them - not where others expect them.

◈ *"What part of God's mandate gives me permission to step in?"*

Leadership Practice: Identify one "off-limits" space you've felt called to influence.

5. Inspire from Knowing, Not Noise

Peter moved from denial to declaration not because of excitement, but because of knowing. Inspiring Presence isn't loud - it's formed.

◈ *"What do I now know of Christ - not just believe - that steadies me?"*

Leadership Practice: Spend time meditating on this question.

---

6. Fortify for Blessing *and* Persecution

True inspiration prepares people not just for favor, but for fire. Leaders who inspire do not romanticize calling - they prepare others for what it will cost.

◈ *"Am I preparing others for pressure, not just promise?"*

Leadership Practice: In your next teaching, mentoring, or message, meditate on this.

---

7. Awaken, Don't Impress

Inspiring Presence is not charisma. It is essence awakening essence. It doesn't entertain - it calls forth.

◈ *"Did I perform, or did I awaken something deeper in the other person?"*

Leadership Practice: Reflect on this about your last three interactions.

✾ Final Charge:

Inspiring Presence is not about hype - it is about holy disruption.
It is what causes nets to be left behind.
It is what turns failures into founders.
It is what enables you to say, *"What I do have, I give you: Rise up and walk."*

The Church doesn't need more personalities. It needs more presence.
The world isn't waiting for excitement. It's waiting for people who know.
Be one who carries life. Who breathes courage. Who stands firm when others fade.

May your presence awaken identity.
May your life stir the divine in others.
May your leadership birth knowing.
And may the world, encountering you, know - not just believe - that Jesus lives.

The Power of Less

A rod in Moses' hand.
A sling in David's grip.
Five loaves, two fish
Just fragments on a dish.
Yet in the hands of purpose,
and under heaven's breath,
less becomes abundance,
and little conquers death.

CHAPTER 7

Leveraging Presence: Designed, not Denied

"I can do all things..."
Philippians 4:13

Introduction

In 1949, in Crosby, Liverpool, a bootless Nigerian football (soccer) team stunned a British crowd by beating their well-equipped opponents 5–2. One commentator remarked on "the shock tactics of the bare-toed Nigerians" who "kept up a terrific pace" and scored great goals. They may have been barefoot, but they were not without presence. Their win didn't just defy expectation - it revealed design.

In athletics, May 25, 1935, is often remembered as *The Day of Days*. On that day, Jesse Owens - son of a sharecropper, grandson of a slave - set three world records and tied a fourth in just 45 minutes at a meet in Ann Arbor, Michigan, during the height of American segregation. He would go on to win four gold medals at the 1936 Berlin Olympics, to the horror of both Adolf Hitler and the segregationists back home.

Owens' performance punched a hole in Nazi and white supremacist ideology - much of which distorted Scripture to justify itself. His triumph was a declaration: **he was God's handiwork - fearfully and wonderfully made and designed, not denied.**

Writing Your Own Story

The enemy's oldest trick is to shift our attention from what we've been given to what we've been denied. This was on full display in Eden. Adam and Eve, though made in the very image of God, were lured away from the truth of their design by the suggestion that something essential had been withheld from them.

Yet of all God's creation, only humanity received the unmatched gift of divine design. Nothing speaks more clearly to our built-in capacity and permission than this: **we were made in the image and likeness of God.** Adam and Eve had not yet begun to tap into the full potential of that design when Satan intervened, cutting short the chance to see what humanity could become simply by walking in dominion.

When we focus on what we've been denied, we delay the writing of our own stories - stories that reveal the power of being. Take George Washington Carver. Britannica describes him as a "revolutionary American agricultural chemist, agronomist, and experimenter." If anyone fit the image in Edwin Markham's 1898 poem *The Man with the Hoe* - the bent figure robbed of purpose - it was Carver. Born into slavery in Missouri, he was kidnapped as an infant by night raiders along with his mother and sister. He was the only one recovered and returned to his master. His father had died before his birth. He grew up without his immediate family.

And yet - he became extraordinary.

After abolition, Carver's intellectual curiosity was evident, and his former master encouraged it. Despite every disadvantage, his achievements in developing new products from peanuts, sweet potatoes, and soybeans surpassed those of many born into privilege. The reason? Though man tried to deny him, **God had designed him** to do extraordinary things with whatever was in his hand.

One defining moment came when Carver, newly freed and seeking education, introduced himself to a woman named Mariah Watkins as "Carver's George" - the way a former slave would signal to whom he had belonged. She replied firmly, *"You are not Carver's George. You are George Carver."*

That moment reshaped his self-concept. The awareness that he was more than the circumstances of his birth would eventually take him to Iowa State University in 1891, where he became the school's first Black student. His bachelor's thesis title speaks volumes about his understanding of the dominion God had embedded in him from creation: *Plants as Modified by Man* (1894)

Leveraging Presence begins not with giftedness, but with the **awareness of design.** George Carver may have already been driven, but the Inspiring and Validating Presence of people like Mariah Watkins helped him stop identifying himself as someone who belonged to the plantation - and start living as someone who belonged to God.

Not only did Carver become known for his inventions and agricultural breakthroughs, but he is also credited with helping save the rural South's economy. And he never pointed to himself as the source of his brilliance. In his own words:

> "The Lord always provides me with life-changing ideas. Not that I am special. The Lord provides everyone with life-changing ideas. These ideas are quite literally a treasure from the Almighty. It is up to each of us, however, to choose and dig for the treasure."

His life is the very picture of Leveraging Presence the act of stewarding divine design into redemptive action. It is also a story of the Validating Presence of those who raised him post-slavery, and the *Inspiring Presence* of those who renamed him, spoke life into him, and empowered him to rise.

I am one among millions born in Third World countries during colonial times, whose parents were the heroes that helped them write their own stories. I spent seven years of elementary school walking barefoot, studying by candlelight, and navigating pit latrines where snakes occasionally lurked in the night. But none of that changed this truth: we were designed, not denied.

On my parents' worn bookshelf were books by and about Booker T. Washington and George Washington Carver - reminders that we could prevail not because of our conditions, but because of our design.

From Victims to Testimony Builders

Leveraging Presence models the power of God's design. To be made in His image is to be both a recipient and a replicator of creative potential. In Genesis 1, God creates *ex nihilo* - out of nothing. In Genesis 2, He forms man *ex material* -from dust.

Both acts reveal who God is and who we are. He creates, then calls us to co-create. He speaks light, then shapes life. He builds a world, then places us in it with a mandate to multiply, steward, and prevail.

In Judges 6:14, God says to Gideon: "Go in this thy might." - charging him to deliver Israel from the Midianites. But while God sees might in Gideon, Gideon is consumed by what he lacks:

"With what shall I save Israel?" (v.15)

God had already answered the question: With your might. But like many of us, Gideon couldn't see the value of what he already carried. He protested what God already knew - that he was the youngest in his family, and that his family was "the poorest in Manasseh." This is the trap religion often sets: It teaches people to keep asking for what God has already given.

The problem is that what God has given often doesn't look like what we were asking for. Gideon was asking for might. God had already said, "Go in this thy might." He had it - it just didn't look like what he expected.

Back in Chapter 1, we explored Radiant Presence, the power of image-bearing essence: like uranium, invisible, weighty, and impossible to ignore. When God sends someone into a space marked by disgrace or danger, their authority is not in their volume - it's in their being.

Leveraging Presence recognizes that essence. It sees what God sees - even before it manifests. The role of the faith community is not only to model Leveraging Presence, but to teach it. It is this kind of presence that transforms "the poorest in Manasseh" -or the forgotten in Compton, the at-risk youth of the South Bronx, the enslaved children like George Washington Carver - from victims into testimony builders.

Leveraging Presence empowers us to write our story the way God intended.

To slay giants with the sling already in our hand - not the sword we think we need. In our work, we carry a sacred responsibility: To help the poor discover their might - the very thing they can leverage in the fight against the Midianites of poverty, oppression, and despair. Ultimately, God's promise to Gideon came down to one thing:

> "Surely I will be with thee, and thou shalt smite the Midianites as one man." (Judges 6:16)

That is the power of Leveraging Presence: Not just knowing that God is with you, but acting on what He has already placed within you.

The Power to Start

A famous Chinese proverb says, *"A journey of a thousand miles begins with a single step."* For biblical figures like Gideon and Moses, the problem wasn't just where to start - it was **with what.** They knew who the enemy was and where to find him.

But they lacked what they assumed was necessary to begin. No army. No influence. No roadmap. Just a call.

Leveraging Presence does not despise "the day of small things" (Zechariah 4:10). It recognizes that we are designed to turn small things into big things - to build consequential legacies with nothing in hand but a seed in the mind and the weight of presence in the soul.

Genesis 1 opens with a void: formless, empty, and dark. The only presence in that space was God. No raw materials. No blueprint. No scaffolding. God began by *being*. Then He spoke. Creation was not outsourced. It began in Him. The One who *was* became the One who *spoke*, and what was spoken became what now is. John 1 takes us even deeper:

> "In the beginning was the Word, and the Word was with God, and the Word was God...All things were made through Him, and without Him nothing was made that was made." (John 1:1–3)

Not only does this locate Jesus at the beginning, it confirms His oneness with the Father. It declares that He *is* God. Earlier in this book, we reflected on the tragedy of ignoring our origin story when trying to understand who we are and what we're capable of. John 1 brings us back there - with Jesus.

And this is exactly what the Great Commission does: It brings us back to the beginning - so we can step forward with clarity and power. "All things were made through Him, and without Him nothing was made that was made."

This can be read two ways:
- That everything was made *by* Him.
- That everything came *out* of Him.

Both are true. There was nothing else to draw from - so all things came from **within** Him. And all things were created *by* Him, because He alone is Creator.

So how does this empower us to start? Because we are made in His image and likeness, we too are designed to **create in voids,** to bring forth in places where nothing yet exists. Things are meant to begin in us.

Secondly, every journey of faith is not a solo trek - it is a journey **with God.**

So we don't only leverage resources - we leverage relationship. We start - not because conditions are perfect, but because we are not alone.

Paul affirms this in Acts 17:

"And He has made from one blood every nation of men to dwell on all the face of the earth, and has determined their preappointed times and the boundaries of their dwellings, so that they should seek the Lord, in the hope that they might grope for Him and find Him, though He is not far from each one of us." (vv.26–27)

Wherever we find ourselves - whatever conditions we are born into - we are never far from God. Paul continues:

"For in Him we live and move and have our being… for we are also His offspring." (v.28)

And so we start. Not because we see the full path. Not because we possess everything we think we need. But because **in Him, we live.** Because **in Him, we move.** Because **in Him, we already are.**

Reclaiming the divinity of our humanity gives us not just vision - but permission.

It gives us the power to start.

The Pole is Long Enough

We start by leveraging what we already know. God has already given - and revealed - more than enough to live by, act on, and teach from for generations.

Yet often, we keep asking for more, even when we've received enough to begin.

As Paul observed, some are "ever learning and never able to come to the knowledge of the truth" (2 Timothy 3:7).

Archimedes famously said: "Give me a place to stand and a lever long enough, and I will move the world." But many spend their lives **measuring the pole** - wondering if it's long enough, doubting its reach, dismissing its potential. No matter the yardstick, it never *seems* sufficient.

Back in Chapter 4, we explored *Alleviating Presence* - not as a ministry of maintenance, but of empowerment. Alleviation is not about keeping people dependent; it's about awakening the reality that they were never helpless to begin with. It refuses to let the hungry see food as destiny, or the hurting see pain as permanent. It opens people's eyes to what's already in their hand. To see that the pole is, in fact, long enough.

In *The Resource Curse*, Syed Mansoob Murshed poses a sobering question:

> "How could nature's bounty turn into a curse? It strains credulity to think that environmental gifts - land, water, forests, minerals, and fuels - could become a curse for nations or peoples richly endowed with them..."

And yet, this paradox is real. Some of the poorest nations on earth are also among the richest in natural resources. Many of these countries have experienced dramatic spiritual revivals and generous waves of global aid. And yet poverty remains. Why?

These contradictions should compel the faith community to ask hard questions: Why isn't the gospel we preach - or the charity we offer - enough to convince people that the pole is long enough? What is missing in our teaching, in our giving, and in the presence we model?

If people never recognize God's design as enough, they remain trapped in someone else's script. *Leveraging Presence* happens when individuals take what appears small, ordinary, or insufficient - and activate it for purpose, truth, and flourishing.

The idea that faith is a pole long enough to move anything isn't poetic embellishment. It's Kingdom logic. Hebrews 11 reminds us that the heroes of faith were not superhuman. They were ordinary people who knew how to leverage what they had:

- **Noah** leveraged obedience and a hammer - and preserved humanity.
- **Sarah** leveraged belief - and carried what was biologically impossible.
- **Abraham** leveraged a promise - and built a nation.
- **Moses** leveraged a rod - and led a people through water.
- **Rahab** leveraged trust and hospitality - and redefined her legacy.

Each of them transformed the ordinary into the extraordinary. They didn't act from surplus. They acted from *trust*. What they held - though seemingly insufficient - became enough because it was surrendered.

That is the power of faith: It is a *lever that moves entire realities.*

And when rooted in the Spirit of God, it's not fueled by fear - but by power, love, and a sound mind (2 Timothy 1:7).

This is what Leveraging Presence looks like:

- **It points the way** - bringing clarity to vision and revealing what is possible, even when no one else sees it.
- **It validates truth** - turning belief into demonstration, and unseen convictions into visible action.
- **It cultivates life** - activating dormant gifts, restoring broken systems, and redeeming people and places for flourishing.

Leveraging Your Influence

Nehemiah is a fitting character to close this book with. His life teaches us that God often places us in seemingly humble or peripheral positions— not as a reflection of our insignificance, but as strategic platforms for Kingdom influence. Influence, after all, is not always about titles or public recognition. It's about access, trust, and timing.

Few exemplify *Leveraging Presence* better than Nehemiah. His job wasn't to lead armies or oversee public works, but to serve as the cupbearer to King Artaxerxes of Persia - tasked with tasting the king's wine to protect him from poisoning. It was a role of great risk, but also great trust. Nehemiah stood daily in the presence of one of the most powerful monarchs on earth - not because of political savvy or noble birth, but because he was faithful, loyal, and unshakably dependable.

Socially, he was still just the king's cupbearer, only slightly above the rest of the Israelites in captivity - distinguished only by where he worked. Rebuilding Jerusalem's walls was not part of his job description. But Nehemiah understood something deeper: his proximity to the throne was not accidental. Even if, to others, he was no more than a house servant, he knew there was a greater purpose.

"I was the king's cupbearer." - Nehemiah 1:11

These words were an acknowledgment of divine placement. For Nehemiah, his role was not merely occupational - it was positional. God had placed him in the palace to do more than sip wine; he was there to steward influence.

Strategy: Favor First

Nehemiah didn't enter the king's chamber with entitlement. He began with prayer - an appeal not to his own eloquence, but to God's favor:

"O Lord, I pray, please let Your ear be attentive... and grant him mercy in the sight of this man." - Nehemiah 1:11

He understood that the assignment before him was far beyond his official mandate. He wasn't a governor - not yet. He didn't own land, command troops, or wield political power. But he had something powerful: the trust of the king - and even more importantly, the settled knowing that God had placed him in that palace for a reason.

So, with humility and clarity, he made the ask:
- **Send me** (Nehemiah 2:5) – Leveraging access
- **Give me letters** (Nehemiah 2:7) – Leveraging authority

- **Grant me timber** (Nehemiah 2:8) – Leveraging resources

Nehemiah didn't confuse his spiritual burden with political rebellion. He didn't attempt to do God's will without God's order. Instead, he respected the chain of command while remaining fully aligned with the God who had placed him there. He understood that **favor is not a shortcut around structure -** it's the power to move within it. He modeled what it means to honor earthly systems without compromising divine allegiance, leveraging divine placement with precision, boldness, and grace.

Field Intelligence Before Vision Casting

When Nehemiah arrived in Jerusalem, he didn't begin with a grand announcement or gather leaders to unveil a master plan. He didn't position himself as a hero sent from the palace. Instead, he moved quietly - by night - inspecting the brokenness in silence:

"I told no one what my God had put in my heart to do at Jerusalem... and the officials did not know where I had gone or what I had done." - Nehemiah 2:12, 16

This is the mark of a true leader: field knowledge before fanfare. Nehemiah understood that influence doesn't come from abstract information - it comes from immersion. He needed to see it for himself, walk through the ruins, feel the dust beneath his feet, and absorb the weight of what had been lost. Only then would his words carry the weight of lived understanding - and only then could he speak with true authority.

Confronting Spiritual Apathy

After assessing the devastation firsthand, Nehemiah gathered the officials, priests, nobles, and the people. Then he spoke:

"You see the distress that we are in, how Jerusalem lies waste... Come and let us build, that we may no longer be a reproach." - Nehemiah 2:17

These words spoken not from assumption, but from lived awareness - cut through the haze of apathy. Nehemiah's voice, shaped by quiet observation and spiritual burden, accomplished what titles and positions alone could not.

And something remarkable happened: the first to rise was not a builder or civic leader, but Eliashib, the high priest (Nehemiah 3:1). Nehemiah's presence - grounded in humility, clarity, and conviction - stirred even the spiritual establishment out of passivity. His leadership reignited what had been dormant for too long.

Strategic Mobilization: Leveraging the Personal

When threats and opposition arose - as they always do - Nehemiah didn't respond with panic. He responded with strategy. He stationed people along the wall, not randomly, but according to their families:

"I set the people according to their families, with their swords, their spears, and their bows."
-Nehemiah 4:13

Why? Because when people are fighting for something personal - their sons, their daughters, their wives, and their homes - they fight differently.

"Do not be afraid of them. Remember the Lord, great and awesome, and fight for your brethren..." - Nehemiah 4:14

This wasn't just organizational brilliance. It was spiritual insight into the power of **relational leverage.** Nehemiah mobilized not through hype, but through heart. He knew that the deepest motivation comes not from abstract ideals, but from personal conviction - from fighting for what matters most.

As Nehemiah Prevailed, So Can You. Nehemiah wasn't a prophet. He wasn't a warrior. He wasn't a king. He was a servant with access, a man with favor, and a leader with clarity. And yet, his presence shifted history. He leveraged presence - in the palace, in the field, in the face of opposition, and before the people. The wall was rebuilt not merely with bricks, but with borrowed authority, divine strategy, and courageous influence. And through it all, Nehemiah never lost sight of who had placed him there:

"The God of heaven Himself will prosper us; therefore we His servants will arise and build."
- Nehemiah 2:20

His life summarizes everything this book has been about.

- He walked in the **weight of being** - anchored in **Presence** (Chapter 1), not performing, but embodying who he was.
- He acted with **Relevance** (Chapter 2), recognizing the need was not theoretical but urgent - a broken city, a disgraced people.
- He began with **Exploration** (Chapter 3), inspecting the ruins before casting vision - seeking wisdom, assessing reality, and grasping complexity.
- He offered **Validation** (Chapter 4), not just by casting vision, but by calling others - priests and people alike - into purpose and participation.
- He practiced redemptive **Alleviation** (Chapter 5), not through pity, but by refus-

ing to normalize brokenness and committing to lift the shame.

- He carried **Inspiration** (Chapter 6), not rooted in hype or status, but in settled knowing. His words stirred courage. His presence awakened resolve.

And when it mattered most, **he leveraged it all** - his position, his access, his relationships, his credibility, his courage, his favor, and his faith.

This is what it means to PREVAIL:
To stand in the full stature of your divine design.
To bring presence into places of need.
To show up as both a reflection of God's image and a conduit of His intention.
Nehemiah prevailed because he knew who he was and who had sent him -
and he walked in the fullness of that knowing.
His presence changed everything.
So must ours.

A Leader and kingdom citizen Blueprint for Leveraging Presence

The Core Premise

Leveraging Presence turns potential into purpose. It is the sacred act of refusing to let lack define us. It sees what's already in the hand, the house, the heart—and activates it. It is not rooted in what is missing, but in what is present. This is not about waiting for resources. It's about awakening resourcefulness.

---

1. Start With Design, Not Denial

Adam and Eve were the only beings made in God's image. But Satan got them to focus on what was withheld instead of what was entrusted. We still fall for the same lie.

◈ *"Are you leading people to focus on what's missing- or what's already been given?"*

Leadership Practice: Make a list of ten things already in your hand - talents, relationships, access, insights. Strategize how one of them can be activated this week.

---

2. Teach People to See Their Might

God told Gideon, "Go in this thy might." But Gideon could only see what he lacked. Many in our communities carry power they don't recognize because they haven't been taught to see it.

◈ *"Do you describe your people as deficient - or designed?"*

Leadership Practice: Audit your language. Reframe how you speak about your team or community. Start naming design and potential, not just needs.

---

3. Activate the Pole in the Hand

Faith is a lever long enough to move anything. But many leaders keep measuring the pole instead of using it. George Washington Carver turned peanuts into legacy. Moses turned a rod into deliverance.

◈ *"What ordinary tool or trait have you discounted?"*

Leadership Practice: Identify one "insufficient" asset in your context. Use it this week - practically, not symbolically - to solve a problem or serve a purpose.

---

4. Move From Victimhood to Testimony

Carver was born into slavery - but walked in design. Gideon was hiding - but was already mighty. God doesn't just rescue us. He repurposes us.

◈ *"Are you helping people rehearse their pain - or rewrite their story?"*

Leadership Practice: Lead a testimony-building session. Don't just collect survival stories. Gather stories of transformation and activation.

---

5. Begin With What You Have

Genesis 1 begins in a void. But God was there - and so creation happened. Nehemiah was a cupbearer, not a governor - but he rebuilt the walls because he understood placement and presence.

◈ *"Are you waiting for more - when Presence says begin now?"*

Leadership Practice: Launch one initiative this month using only the resources you currently have. Don't wait. Start. Let God multiply.

---

6. Leverage Personal Access for Public Good

Nehemiah leveraged his access to the king for the good of a broken city. He used favor, letters, timber, and trust - not for personal gain, but for collective restoration.

◈ *"What doors can you open for others that they can't open for themselves?"*

Leadership Practice: Map your influence. Identify three people in your network whose access could benefit someone else. Make an intentional connection this week.

---

7. Strategize for Ownership, Not Hype

Nehemiah didn't stir a crowd with grand speeches. He walked the ruins quietly, then mobilized people around their families and homes - what mattered most.

◈ *"Are you organizing for applause - or for transformation and ownership?"*

Leadership Practice: Restructure one team or project based on relational lines - families, neighborhoods, shared values - and reassign responsibility accordingly.

Final Charge

Leveraging Presence doesn't wait for perfection. It begins with awareness, ignites with trust, and multiplies through faith. It points to the pole and says: *"It's long enough."* It sees Nehemiahs in wine cellars, Gideons in hiding, Carvers in cotton fields - and calls them to rise.

Stop asking what's missing.
Look at what's in your hand.

Look at who's in your house.

Look at what God already placed in your soul.

Leverage it.

That is **Leveraging Presence**.

That is **Kingdom leadership**.

That is how we **PREVAIL**.

The greatest threat to the
kingdom of darkness
is a believer who awakens the
divinity of their humanity - and
that of those they raise or lead -
draining the swamp from which
the enemy recruits us to wage war
against ourselves.

EPILOGUE
The Call to Prevail

"You have made them a little lower than the angels and crowned them with glory and honor. You made them rulers over the works of your hands..."

Psalm 8:5–6

You Were Created to Prevail

You were not created to merely survive, to seek safety in obscurity, or to earn worth through spiritual performance. You were made in the image and likeness of God - crowned with glory and honor, designed to carry weight, and sent into the world with divine authority.

This book has not been about motivation. It has been about reawakening. It is a summons to recover the divine blueprint embedded in your being. To reclaim the dominion you were designed to steward. To move from reacting to reforming, from waiting to wielding.

To prevail is not to overpower. It is to embody the full measure of your design - to show up in every sphere of life with presence that points the way, validates truth, and cultivates life.

This is not just personal. It is generational. It is institutional. It is cultural. The world awaits not more content, but more carriers of glory.

The Tyranny of Helplessness Ends Here

Many people of faith have learned to spiritualize powerlessness. We've confused humility with invisibility, silence with holiness, and waiting with obedience - even when God has already spoken.

But the world is not transformed by good intentions or hidden wisdom. Scripture warns: "The poor man's wisdom is despised, and his words are not heard." (Ecclesiastes 9:16)

This is not a reason to withdraw. It is a call to show up

- Not louder, but **deeper.**
- Not just anointed, but **anchored**.
- Not just compassionate, but **competent**.
- Not just praying, but **prevailing**.

The tyranny of helplessness ends when we remember that we have always had something in our hand - presence, identity, and design. The pole is long enough. The rod still parts the waters. The church still has what the world needs.

A New Breed of Builders

The world does not need more performers. It needs presence-bearers. Not just voices that stir emotions - but lives that **awaken essence**.

Not just projects that manage crisis - but people who activate destiny.

PREVAIL calls forth a new generation of:

- **Designers** who don't just make things look good, but make things work redemptively.
- **Entrepreneurs** who don't just generate profit, but create pathways to dignity.
- **Educators** who don't just transfer information, but form identity.
- **Faith leaders** who aren't trapped in sanctuary systems, but become architects of societal wholeness.
- **Parents** who disciple destiny, not just discipline behavior.
- **Young people** who trade shame and striving for sonship and stewardship.

Builders. Reformers. Restorers. Strategists. Sons and daughters walking in divine design.

PREVAIL Is More Than a Model It is a framework for formation. A compass for leaders. A mirror for maturity. A tool for discernment. A challenge to systems. A blueprint for kingdom impact.

- **Presence** – Reclaim your essence.
- **Relevance** – Step into real-world significance.
- **Exploration** – Let curiosity become a holy habit.
- **Validation** – Speak to the image of God in others.
- **Alleviation** – Bind wounds, break cycles, build lives.
- **Inspiration** – Breathe life, not just hype.
- **Leverage** – Activate what's already in your hand.

These aren't concepts to admire. They are dimensions to embody.

So let this book not be a conclusion, but a commissioning. Let it call you to journey back to the beginning not alone, but with Jesus, back to the garden, the radiant cradle, the image, to the breath and forward to your assignment in this hour and this generation to God's glory.

Final Charge:

Let your presence be weighty.
Let your voice carry clarity.
Let your leadership be redemptive.
Let your impact be undeniable.
Let your life be a holy disruption in a world of despair.
Let the Church be dangerous again.
Not because it dominates, but because it dignifies.
Not because it shouts, but because it awakens.

Don't just read this book.
Don't just believe.
Don't just build.
Reclaim your presence.
Redefine your metrics.
Reshape the world.

Prevail.

PART TWO

Self-Assessment Tools for Prevailing Leadership and Kingdom Citizenship

The Scorecards

Preface

A Reflective Journey Toward Prevailing Leadership and Kingdom Citizenship

Throughout *Prevail: Reclaiming the Divinity of Our Humanity*, you've explored seven dimensions of redemptive leadership and kingdom citizenship: Presence, Relevance, Exploration, Validation, Alleviation, Inspiration, and Leveraging. These dimensions are not theoretical ideals - they are practical expressions of the divine image within you, meant to shape your life, your leadership, and your impact.

To help you apply these principles meaningfully, the following Prevail Scorecards have been collected into one cohesive section. They are not exams. They are invitations to reflect - to pause, assess, and realign with your identity and purpose.

Each scorecard corresponds to a chapter and is designed to:

- Help you *assess your current alignment* with each leadership principle,
- Prompt *honest reflection* on areas of growth,
- Encourage *intentional action* toward becoming who you were created to be.

How to Use These Scorecards

1. **Read and Reflect**
   After completing the book - or any specific chapter - revisit the corresponding scorecard. Don't rush through it. Let each question probe the posture of your heart and the habits of your life.

2. **Rate Honestly**
   Each scorecard invites you to rate yourself on a scale from 1 (Rarely) to 5 (Consistently). There are no perfect scores. The goal is not performance, but awareness and alignment.

3. **Respond Intentionally**
   Each scorecard includes reflection prompts and action steps. Use them to journal, set leadership goals, or initiate change in the environments you influence.

4. **Revisit Periodically**

   Transformation is a process. These score-cards are designed to be used **again and again** as part of your ongoing spiritual and leadership development. Revisit them quarterly, annually, or during seasons of transition.

5. **Share in Community**

   If you're journeying through *Prevail* in a group or leadership cohort, these score-cards can spark vulnerable, catalytic conversations. Use them to **mentor others, lead team reflections**, or support discipleship and leadership development.

You were created to prevail.

These scorecards are simply tools to help you do so with *clarity, conviction, and the courage to walk in the fullness of who you are* - not just in theory, but in presence, purpose, and practice.

Presence Scorecard

This scorecard is designed to help you assess your alignment with the principle of *Presence* as outlined in Chapter 1. Presence is the weight of being, rooted in the divine identity of being created in God's image. Reflect honestly on each question to evaluate how you embody this foundational dimension of impact. For each question, rate yourself on a scale of 1 to 5, where 1 represents "Rarely" and 5 represents "Consistently." After scoring, calculate your total to gauge your current state of Presence.

Scorecard Questions

1. **Identity Awareness**: Do you consciously recognize and embrace your identity as created in the image and likeness of God, carrying inherent worth and divine purpose?
   ◦ (1 = Rarely, 5 = Consistently)
   ◦ *Reflection*: Consider how often you reflect on your divine essence versus external labels or limitations.

2. **Weight of Being**: Do you bring the full weight of your authentic self into your interactions, environments, and responsibilities, without shrinking or conforming to external pressures?
   ◦ (1 = Rarely, 5 = Consistently)
   ◦ *Reflection*: Think about moments when you felt fully present versus times you felt absent or diminished.
3. **Radiant Impact**: Are you aware of how your presence affects others and your surroundings, radiating God's goodness, hope, or transformation without needing to act or speak?
   ◦ (1 = Rarely, 5 = Consistently)
   ◦ *Reflection*: Recall instances where your mere presence shifted an atmosphere or influenced someone positively.
4. **Settled Knowing**: Do you operate from a settled confidence in your divine calling and capacity, trusting that you are equipped to fulfill your purpose regardless of challenges?
   ◦ (1 = Rarely, 5 = Consistently)

- ◦ *Reflection*: Assess whether you doubt your abilities or lean into the assurance of your God-given design.

5. **Scriptural Anchoring**: Do you draw strength and clarity from Scripture (e.g., Genesis 1, Psalm 16:11, or 2 Timothy 1:7) to reinforce your understanding of your identity and presence?
   - ◦ (1 = Rarely, 5 = Consistently)
   - ◦ *Reflection*: Evaluate how often you turn to biblical truths to ground your sense of self.

Scoring

- Add your scores for each question (range: 5–25).
- **Interpretation**:
  - ◦ **5–10**: Your presence may be underdeveloped, possibly due to doubts or disconnection from your divine identity. Revisit Chapter 1 and focus on embracing your essence as God's image-bearer.
  - ◦ **11–17**: You show moments of presence but may waver under pressure. Reflect

on areas where you can lean more fully into your settled knowing.

○ **18–25**: You consistently embody a weighty, radiant presence. Continue cultivating this strength to influence your environments for God's glory.

Reflection and Action

- **Key Insight**: Presence is not about doing but being. Your divine essence, rooted in God's image, is a radiant gift that transforms spaces before you act.
- **Action Step**: This week, practice intentional presence in one specific environment (e.g., workplace, home, community). Before entering, pause to affirm, "I am created in God's image, and my presence carries His weight." Notice how this shifts your confidence and impact.
- **Journal Prompt**: Write about a moment when you felt fully present and radiant. What enabled that moment, and how can you replicate it?

Relevant Presence Scorecard

This scorecard is designed to help you assess your alignment with the principle of *Relevant Presence* as outlined in Chapter 2. Relevant Presence is the practical, impactful expression of your divine essence in real-world contexts, bringing transformation and influence like the "salt of the earth" (Matthew 5:13). Reflect honestly on each question to evaluate how you embody this dimension of impact. For each question, rate yourself on a scale of 1 to 5, where 1 represents "Rarely" and 5 represents "Consistently." After scoring, calculate your total to gauge your current state of Relevant Presence.

Scorecard Questions

1. **Engagement with Real-World Needs**: Do you actively seek to understand and address the practical needs and challenges in your environment (e.g., workplace, community, or marketplace) with your God-given gifts?
   - (1 = Rarely, 5 = Consistently)

- *Reflection*: Consider whether you engage with real-world issues or retreat into spiritual or personal comfort zones.
2. **Full Expression of Divine Design**: Do you employ the full range of your divine faculties—rationality, creativity, moral agency, relational capacity, and purposeful dominion—in your daily interactions and responsibilities?
   - (1 = Rarely, 5 = Consistently)
   - *Reflection*: Think about whether you limit yourself to moral or spiritual expressions or fully utilize your God-given capacities.
3. **Courageous Presence in Difficult Spaces**: Do you bring your authentic, God-anchored presence into challenging or secular environments (e.g., boardrooms, classrooms, or civic spaces) without fear or compromise?
   - (1 = Rarely, 5 = Consistently)
   - *Reflection*: Recall moments when you stepped into tough spaces with confidence versus times you avoided or conformed.
4. **Impactful Influence**: Does your presence naturally influence and transform your sur-

roundings (e.g., preserving, purifying, or enhancing like salt), making a tangible difference without needing to force attention?

- (1 = Rarely, 5 = Consistently)
- *Reflection*: Assess whether your presence shifts atmospheres or goes unnoticed in key settings.

5. **Alignment with Divine Mandate**: Do you view your work, leadership, or contributions as part of God's commission to "be fruitful, multiply, and have dominion" (Genesis 1:28), rather than as secular or secondary to spiritual pursuits?

- (1 = Rarely, 5 = Consistently)
- *Reflection*: Evaluate how often you connect your daily efforts to God's redemptive purpose for the world.

Scoring

- Add your scores for each question (range: 5–25).
- **Interpretation**:
  - **5–10**: Your relevant presence may be limited, possibly due to fear, disconnection from your divine mandate, or over-

emphasis on spiritual isolation. Revisit Chapter 2 to embrace your role as salt in the world.

- ○ **11–17**: You show moments of relevant presence but may hesitate in fully engaging complex or secular spaces. Reflect on how to bring more of your essence into these arenas.
- ○ **18–25**: You consistently embody a relevant, impactful presence that transforms environments. Continue stewarding this influence to advance God's Kingdom.

Reflection and Action

- **Key Insight**: Relevant Presence is about bringing the full weight of your divine essence into real-world spaces, transforming them through practical, courageous, and purposeful engagement.
- **Action Step**: This week, identify one specific context (e.g., a workplace meeting, community event, or civic issue) where you can bring your God-given essence more fully. Prepare by affirming, "I am the salt of

the earth, called to bring life and redemption here." Observe the impact.

- **Journal Prompt**: Write about a time when your presence made a tangible difference in a challenging environment. What enabled your relevance, and how can you cultivate this consistently?

Exploring Presence Scorecard

This scorecard is designed to help you assess your alignment with the principle of *Exploring Presence* as outlined in Chapter 3. Exploring Presence is the courageous, curious engagement with God's creation and truth, driven by a divine impulse to seek, ask, and discover, as exemplified by Jesus in the temple and scientists like Dr. Jennifer Wiseman. Reflect honestly on each question to evaluate how you embody this dimension of impact. For each question, rate yourself on a scale of 1 to 5, where 1 represents "Rarely" and 5 represents "Consistently." After scoring, calculate your total to gauge your current state of Exploring Presence.

Scorecard Questions

1. **Curiosity in Action**: Do you actively seek new knowledge or experiences (e.g., through study, cross-cultural engagement, or exploration of creation) to deepen your understanding of God and His world?
   ◦ (1 = Rarely, 5 = Consistently)

- *Reflection*: Consider whether you pursue learning beyond your comfort zone or stick to familiar knowledge.

2. **Overcoming Fear and Pride**: Do you confront fear or pride that hinders exploration, stepping into the unknown with faith rather than clinging to certainty or cultural/traditional identities?
   - (1 = Rarely, 5 = Consistently)
   - *Reflection*: Think about times you avoided exploration due to fear of failure or pride in "knowing enough" versus moments you embraced the unknown.

3. **Engagement with Creation**: Do you view scientific discovery, innovation, or exploration of the natural world as opportunities to encounter God's glory, integrating faith with inquiry?
   - (1 = Rarely, 5 = Consistently)
   - *Reflection*: Assess whether you see exploration (e.g., science, art, or travel) as a sacred act or separate from faith.

4. **Intentional Inquiry**: Do you ask questions, listen actively, and engage with others' perspectives (as Jesus did in the temple) to grow

in wisdom and bring redemptive presence to your environments?

- ◦ (1 = Rarely, 5 = Consistently)
- ◦ *Reflection*: Recall instances where your questions or openness led to insight versus times you stayed passive.

5. **Alignment with Divine Mandate**: Do you see exploration as part of your divine commission to "be fruitful, multiply, and subdue the earth" (Genesis 1:28), using curiosity to co-labor with God in bringing order and beauty?

- ◦ (1 = Rarely, 5 = Consistently)
- ◦ *Reflection*: Evaluate how often you connect your curiosity to God's call to steward and expand His creation.

Scoring

- Add your scores for each question (range: 5–25).
- **Interpretation**:
  - ◦ **5–10**: Your exploring presence may be underdeveloped, possibly due to fear, pride, or a view of curiosity as unspiritu-

al. Revisit Chapter 3 to embrace exploration as a divine invitation.

- ◦ **11–17**: You show moments of curious engagement but may hesitate when faced with uncertainty or institutional resistance. Reflect on how to cultivate bolder inquiry.
- ◦ **18–25**: You consistently embody an exploring presence, seeking truth and wisdom with reverence. Continue using curiosity to advance God's Kingdom.

Reflection and Action

- **Key Insight**: Exploring Presence is about embracing curiosity as a sacred tool to uncover God's truth and glory, enabling you to bring redemptive impact through informed, courageous engagement.
- **Action Step**: This week, pursue one act of exploration (e.g., read a book outside your expertise, engage in a cross-cultural conversation, or study a natural phenomenon). Before starting, pray, "God, reveal Your glory through this discovery." Note how it shapes your presence.

- **Journal Prompt**: Write about a time when curiosity led you to a deeper understanding of God or His creation. What barriers did you overcome, and how can you nurture this posture daily?

Validating Presence Scorecard

This scorecard is designed to help you assess your alignment with the principle of *Validating Presence* as outlined in Chapter 4. Validating Presence is the act of seeing and awakening the divine potential in others, affirming their God-given capacity and calling them into transformation, as Jesus did with the Samaritan woman. Reflect honestly on each question to evaluate how you embody this dimension of impact. For each question, rate yourself on a scale of 1 to 5, where 1 represents "Rarely" and 5 represents "Consistently." After scoring, calculate your total to gauge your current state of Validating Presence.

Scorecard Questions

1. **Seeing Divine Potential**: Do you intentionally look beyond people's circumstances, labels, or brokenness to recognize and affirm their God-given identity and capacity as image-bearers?
   ◦ (1 = Rarely, 5 = Consistently)

- *Reflection*: Consider whether you focus on others' limitations or actively see their divine essence and potential.

2. **Encouraging Transformation**: Do you engage with others in ways that inspire and equip them to move beyond dependency or brokenness toward wholeness and purpose, rather than merely comforting them?
   - (1 = Rarely, 5 = Consistently)
   - *Reflection*: Think about whether your interactions maintain people's current state or challenge them to grow into their calling.

3. **Challenging Systems of Dependency**: Do you advocate for or create environments (e.g., in ministry, work, or community) that prioritize transformation and empowerment over perpetuating need or control?
   - (1 = Rarely, 5 = Consistently)
   - *Reflection*: Assess whether you support systems that release people into their potential or inadvertently sustain dependency.

4. **Engaging with Respect and Depth**: Do you interact with others, especially those marginalized or overlooked, with respect,

engaging them in meaningful dialogue that honors their intelligence and capacity, as Jesus did with the Samaritan woman?

- ◦ (1 = Rarely, 5 = Consistently)
- ◦ *Reflection*: Recall moments when you validated someone's worth through deep engagement versus times you dismissed or overlooked them.

5. **Alignment with Divine Mandate**: Do you view your role in others' lives as part of God's commission to "be fruitful, multiply, and have dominion" (Genesis 1:28), calling them into their God-given responsibility and authority?

- ◦ (1 = Rarely, 5 = Consistently)
- ◦ *Reflection*: Evaluate how often you connect your efforts to empower others with God's validating act of entrusting humanity with dominion.

Scoring

- Add your scores for each question (range: 5–25).
- **Interpretation**:
  - ◦ **5–10**: Your validating presence may be underdeveloped, possibly due to a focus

on comfort or alignment with systems that maintain dependency. Revisit Chapter 4 to embrace validation as awakening potential.

- ○ **11–17**: You show moments of validating presence but may hesitate to challenge dependency or fully engage others' potential. Reflect on how to foster transformation more boldly.

- ○ **18–25**: You consistently embody a validating presence, releasing others into their divine calling. Continue cultivating environments that empower and send.

Reflection and Action

- **Key Insight**: Validating Presence sees beyond brokenness to awaken the divine potential in others, transforming them from dependents into co-laborers in God's Kingdom.

- **Action Step**: This week, identify one person in your sphere (e.g., a colleague, friend, or community member) who seems overlooked or stuck. Engage them in a meaningful conversation, affirming their worth and potential with words like, "I see God's strength in you." Observe how this shifts their perspective.

- **Journal Prompt**: Write about a time when someone validated your potential, sparking transformation. How did it feel, and how can you replicate that impact for someone else?

Alleviating Presence Scorecard

This scorecard is designed to help you assess your alignment with the principle of *Alleviating Presence* as outlined in Chapter 5. Alleviating Presence is the intentional, redemptive act of addressing suffering with proximity, participation, and promise, as exemplified by the Good Samaritan and Jesus' holistic compassion. Reflect honestly on each question to evaluate how you embody this dimension of impact. For each question, rate yourself on a scale of 1 to 5, where 1 represents "Rarely" and 5 represents "Consistently." After scoring, calculate your total to gauge your current state of Alleviating Presence.

Scorecard Questions

1. **Proximity to Suffering**: Do you draw near to those who are hurting, engaging directly with their pain through listening, presence, or tangible support, rather than passing by or outsourcing care?
   ◦ (1 = Rarely, 5 = Consistently)

- *Reflection*: Consider whether you actively step into others' suffering or remain at a distance due to discomfort or convenience.

2. **Participatory Action**: Do you give your time, resources, or influence to help restore others, acting with intentionality to address both immediate needs and long-term healing, as the Samaritan did?
   - (1 = Rarely, 5 = Consistently)
   - *Reflection*: Think about whether your help is reactive and fleeting or purposeful and sustained, contributing to others' restoration.

3. **Commitment to Promise**: Do you follow through on your efforts to alleviate suffering, staying invested in others' healing or empowerment, even when it requires ongoing effort or sacrifice?
   - (1 = Rarely, 5 = Consistently)
   - *Reflection*: Assess whether you walk away after initial help or create systems of care that ensure lasting impact.

4. **Discerning Redemptive Love**: Do you offer help guided by wisdom and purpose, distinguishing between enabling dependen-

cy and empowering transformation, as Peter did in Acts 3?

- ○ (1 = Rarely, 5 = Consistently)
- ○ *Reflection*: Recall moments when your compassion led to empowerment versus times it may have perpetuated need.

5. **Holistic Restoration**: Do you address people's needs comprehensively—body, soul, and spirit—aligning with Jesus' model of meeting both physical and spiritual hunger, as in Matthew 14?

- ○ (1 = Rarely, 5 = Consistently)
- ○ *Reflection*: Evaluate how often you consider the whole person in your acts of mercy, ensuring your help restores dignity and purpose.

Scoring

- Add your scores for each question (range: 5–25).
- **Interpretation**:
  - ○ **5–10**: Your alleviating presence may be underdeveloped, possibly due to hesitation, lack of discernment, or limited engagement with suffering. Revisit

Chapter 5 to embrace the Samaritan's standard of redemptive mercy.

○ **11–17**: You show moments of alleviating presence but may struggle with consistency or holistic impact. Reflect on how to deepen your proximity, participation, and promise.

○ **18–25**: You consistently embody an alleviating presence, transforming lives through intentional, redemptive love. Continue building systems that empower and restore.

Reflection and Action

- **Key Insight**: Alleviating Presence is love that labors to restore, acting with proximity, participation, and promise to ensure suffering is not just relieved but uprooted, affirming that every person is important to God.

- **Action Step**: This week, identify one person or situation where you can practice alleviating presence (e.g., mentoring someone in need, supporting a community initiative, or addressing a specific hurt). Act with proximity (engage directly), participation (give meaningfully), and promise (commit to follow through). Note the impact.

- **Journal Prompt**: Write about a time when someone's intentional help transformed your life or perspective. What made their presence redemptive, and how can you mirror that in your actions?

Inspiring Presence scorecard

This scorecard is designed to help you assess your alignment with the principle of *Inspiring Presence* as outlined in Chapter 6. Inspiring Presence is the courageous, transformative act of awakening others' divine essence - moving them from belief to knowing. It is not about charisma or hype, but about living from identity and calling out that same truth in others. Reflect honestly on each question to evaluate how you embody this dimension of impact. For each question, rate yourself on a scale of 1 to 5, where 1 represents "Rarely" and 5 represents "Consistently." After scoring, calculate your total to gauge your current state of Inspiring Presence.

Scorecard Questions

1. Confidence in Divine Identity

Do you carry a settled knowing of your identity as an image-bearer, stepping boldly into challenging spaces (like David in the valley) without being deterred by external limitations or voices like Eliab's?

- (1 = Rarely, 5 = Consistently)
- **Reflection:** Consider whether you shrink from opportunities due to a perceived lack of qualifications or confidently act from your God-given essence.

2. Awakening Others' Potential

Do you inspire others by calling forth their divine potential, encouraging them to move beyond doubt or fear into a deeper knowing of their capacity, as Jesus did with Peter?

- (1 = Rarely, 5 = Consistently)
- **Reflection:** Think about whether your presence stirs others to action and transformation or merely offers temporary encouragement.

3. Balancing Affective and Effective Presence

Do you combine compassion and decisive action in your interactions, inspiring others through both emotional connection and practical impact, avoiding the trap of being only affective or only effective?

- (1 = Rarely, 5 = Consistently)
- **Reflection:** Assess whether your inspiration is rooted in fleeting emotion or fosters lasting change through truth and grace.

4. Persistence Amid Opposition

Do you persist in bringing your presence to spaces where you're told you don't belong, trusting in the "Gift of Gifts" (God's image) to overcome resistance, as David did despite Eliab's rebuke?
- (1 = Rarely, 5 = Consistently)
- **Reflection:** Recall moments when you pushed through opposition to inspire versus times you retreated due to criticism or doubt.

5. Formation Over Excitement

Is your influence rooted in identity that has been shaped by truth, tested by hardship, and formed over time - like Peter's journey from denial to declaration?
- (1 = Rarely, 5 = Consistently)
- **Reflection:** Evaluate whether you lead from the excitement of moments or the

maturity of formation that inspires others to know - not just believe.

Scoring

- Add your scores for each question (**range: 5–25**).

Interpretation:

- **5–10:** Your inspiring presence may be underdeveloped, possibly due to self-doubt, reliance on external validation, or a focus on temporary enthusiasm. Revisit Chapter 6 to root your presence in the Gift of Gifts.
- **11–17:** You show moments of inspiring presence but may waver under opposition or lean too heavily on emotion or action alone. Reflect on how to deepen your impact through essence-driven inspiration.
- **18–25:** You consistently embody an inspiring presence, awakening others' divine potential with courage and clarity. Continue fostering transformation that moves people from belief to knowing.

Reflection and Action

- **Key Insight:** Inspiring Presence is not about charisma or hype but about awakening the divine essence in others, moving them from doubt to knowing through a presence rooted in God's image and mandate.
- **Action Step:** This week, identify one person or group facing a "Goliath" (e.g., fear, doubt, or a challenge). Speak to their potential, saying something like, *"I see God's strength in you to overcome this."* Follow up to encourage their progress. Note the impact.
- **Journal Prompt:** Write about a time when someone's presence inspired you to act beyond your perceived limits. What made their influence transformative, and how can you cultivate that same presence for others?

Leveraging Presence Scorecard

This scorecard is designed to help you assess your alignment with the principle of *Leveraging Presence* as outlined in Chapter 7. Leveraging Presence is the strategic and faithful use of your God-given design, position, and resources to enact redemptive change, as exemplified by George Washington Carver and Nehemiah. Reflect honestly on each question to evaluate how you embody this dimension of impact. For each question, rate yourself on a scale of 1 to 5, where 1 represents "Rarely" and 5 represents "Consistently." After scoring, calculate your total to gauge your current state of Leveraging Presence.

Scorecard Questions

1. **Awareness of Divine Design**: Do you focus on the capacity and permission God has given you as His image-bearer, choosing to see your divine design rather than what you feel denied, as Carver did despite his circumstances?

- ○ (1 = Rarely, 5 = Consistently)
- ○ *Reflection*: Consider whether you dwell on limitations or actively recognize the "might" God has placed within you to prevail.

2. **Strategic Use of Position**: Do you leverage your current role, access, or influence (however small) to advance God's purposes, as Nehemiah did as a cupbearer, rather than waiting for a "better" opportunity?
   - ○ (1 = Rarely, 5 = Consistently)
   - ○ *Reflection*: Think about whether you use your existing platform strategically or hesitate because it feels insignificant.

3. **Action from Small Beginnings**: Do you take steps to act on what you have in hand, trusting that God can transform small things into significant outcomes, as Gideon was called to do with his unrecognized might?
   - ○ (1 = Rarely, 5 = Consistently)
   - ○ *Reflection*: Assess whether you start with what's available (e.g., ideas, relationships, skills) or delay due to perceived lack.

4. **Grounded Preparation**: Do you prepare for action through prayer, observation, and

immersion in the reality of a need (like Nehemiah's nighttime inspection), ensuring your influence is informed and authentic?

- ◦ (1 = Rarely, 5 = Consistently)
- ◦ *Reflection*: Recall moments when you acted with informed conviction versus times you moved impulsively or not at all.

5. **Mobilizing Others**: Do you inspire and organize others to join in redemptive work, leveraging personal connections and shared purpose, as Nehemiah did by rallying families to rebuild the wall?

- ◦ (1 = Rarely, 5 = Consistently)
- ◦ *Reflection*: Evaluate how often you engage others' strengths to amplify impact versus working alone or disengaging.

Scoring

- Add your scores for each question (range: 5–25).
- **Interpretation**:
  - ◦ **5–10**: Your leveraging presence may be underdeveloped, possibly due to a focus on lack, hesitation to act, or failure to

see your position as strategic. Revisit Chapter 7 to embrace your divine design as enough to start.

- ○ **11–17**: You show moments of leveraging presence but may struggle with consistency or confidence in small beginnings. Reflect on how to act boldly with what's in your hand.
- ○ **18–25**: You consistently embody a leveraging presence, transforming opportunities and resources into redemptive impact. Continue stewarding your influence with wisdom and faith.

Reflection and Action

- **Key Insight**: Leveraging Presence is about recognizing and activating the divine design, position, and resources God has given you to write a redemptive story, turning small beginnings into lasting change.
- **Action Step**: This week, identify one area where you can leverage your presence (e.g., a skill, relationship, or role) to address a need or opportunity. Take one concrete step, like Nehemiah's prayer or inspection,

and affirm, "God has placed this in my hand for a purpose." Track the outcome.

- **Journal Prompt**: Write about a time when you or someone else turned a small resource or opportunity into significant impact. What enabled that leverage, and how can you apply that principle now?

About the author

Dr. Noah Manyika is a visionary thought leader, author, and dynamic speaker with over 30 years of experience in transformational community development and global social entrepreneurship. A Fulbright Scholar and former Senior Fellow with the Sagamore Institute, he holds a Master of Science in Foreign Service from Georgetown University's prestigious School of Foreign Service, where he studied under former U.S. Secretary of State Madeleine Albright. He also holds a BA in Journalism and Political Science from Romania's Academia Ștefan Gheorghiu, earned during the Cold War, a PhD in Christian Leadership from Vision University, and a Doctorate in Transformational Leadership from Bakke Graduate University.

His leadership journey began in Zimbabwe, where he held executive business and ministerial leadership roles from 1987 to 1994. After relocating to Charlotte, North Carolina, he founded Nexus Global Serve/Nexus Ministries, leading transformative programs and public-private partnerships to serve inner-city communities. His visionary projects included the Charlotte Children's Scholarship Fund, Brookstone School, and the Charlotte Empowerment Zone. In Zimbabwe, his One Tribe Problem Solvers Clubs facilitated partnerships with civic, business, and government organizations, earning him an invitation to advise the cabinet-level Organ for National Healing, Reconciliation, and Integration during a pivotal period in the nation's history.

Dr. Manyika has served on the Affordable Housing Cabinet established by the late developer John Crosland, Queens University's Advisory Board, and multiple boards, including United World Missions, Harvests of Hope, and Above and Beyond. He is the former Africa Chairman of Kalahari Capital Partners, an investment platform that combined new business creation with short-term trade finance projects in emerging markets.

A candidate in Zimbabwe's 2018 presidential elections, his diverse personal and professional experiences across Africa, Eastern Europe, and the United States have provided him with unique multidisciplinary insights into leadership, global engagement, and transformational change.

Currently, as the founder and CEO of Kitchen Copilot Inc., Dr. Manyika continues to innovate at the intersection of faith and technology, creating tools designed to empower families and communities. He is the author of *The Challenge of Leadership: Is There Not a Cause?*, widely utilized in Bible schools, churches, and leadership training programs, and *Redeeming Sundar: Faith and Innovation in the Age of AI.*

Rooted deeply in faith, Dr. Manyika exemplifies his commitment to God's redemptive vision in both his professional and personal life. He shares this journey with his wife, Phillis, and their three children.

In his latest book, *Prevail: Reclaiming the Divinity of Our Humanity,* Dr. Manyika presents a transformative framework for redemptive Kingdom citizenship and leadership. Using the acronym PREVAIL- Presence, Relevance, Exploring, Validating, Alleviating, Inspiring, and Leveraging - he outlines seven dimensions of impact rooted in our identity as image-bearers of God. When fully lived, these principles offer a structured path to meaningful influence and the fulfillment of humanity's redemptive purpose.

This book, along with Dr. Manyika's broader body of work featured on The Issachar Coll3ctive platform (Issacharcoll3ctive.com), serves as a rallying cry for Christian leaders, social innovators, and believers everywhere to engage the world with courage, creativity, and conviction.

www.ingramcontent.com/pod-product-compliance
Lightning Source LLC
Chambersburg PA
CBHW071330210326
41597CB00015B/1399